Capitol City Fireman

Jake Rixner

Copyright © 2010 by Jake Rixner
First Edition – June 2010

Photographs, #s 8, 12, 13, 14, 15, 16, 17, 18, 19, 20, 21, 22, 23, 24 courtesy of Richmond Times Dispatch. Photograph #28 courtesy of Curtis Phillips.

ISBN
978-1-77067-128-7 (Hardcover)
978-1-77067-129-4 (Paperback)
978-1-77067-130-0 (eBook)

All rights reserved.

The events presented in this book are true, but some of the names were changed to protect the innocent and the guilty.

No part of this publication may be reproduced in any form, or by any means, electronic or mechanical, including photocopying, recording, or any information browsing, storage, or retrieval system, without permission in writing from the publisher.

Published by:

FriesenPress
Suite 300 – 777 Fort Street
Victoria, BC, Canada V8W 1G9

www.friesenpress.com

For information on bulk orders contact:
info@friesenpress.com or fax 1-888-376-7026

Distributed to the trade by The Ingram Book Company

TABLE OF CONTENTS

- Acknowledgments - v
- Forward -................................. vii

- Chapter 1 -
CAPITOL CITY FIREMAN 1

- Chapter 2 -
The Early Years 7

- Chapter 3 -
P.G. COUNTY 17

- Chapter 4 -
THE BIG MOVE....................... 23

- Chapter 5 -
Testing for a Job 31

- Chapter 6 -
First Day in the House................. 37

- Chapter 7 -
The Captain 47

- Chapter 8 -
Fire Fighting Super Stars 59

- Chapter 9 -
Big Fires............................ 69

- Chapter 10 -
Refugee Truck Companies 81

- Chapter 11 -
New Blood 89

- Chapter 12 -
Acting Lieutenant **113**

- Chapter 13 -
Fun with the Cops **123**

- Chapter 14 -
D.C. Fire Department Calls. **129**

- Chapter 15 -
Back in Richmond **149**

- Chapter 16 -
Jackson Ward **157**

- Chapter 17 -
Guardian Angel **165**

- Chapter 18 -
Studying for Promotion **173**

- Chapter 19 -
People Trapped.........Not **185**

- Chapter 20 -
Lieutenants Bars, Are They Worth It?. . . . **191**

- Chapter 21 -
Promotion Day. **203**

- Acknowledgments -

Some of the names in this book were changed to protect the innocent, and the guilty.

I would like to dedicate this book to every fireman that's ever risked his life to save another human being, and in particular the three men that most inspired my life. First is my father who taught me about being a man, overcoming fear, and getting the job done. I wish he was alive to see me graduate first in my class at the Richmond, Virginia Fire Bureau Recruit class in 1982. Second was John "Sully" Sullivan a tough, street smart fireman who had reflexes as quick as a cat, and taught me the tricks of the trade inside burning buildings. Third was Peter B. Lund. I met Pete in the Bronx in 1990 during a ride-along when he was at Rescue Co. 3. We shared the same views about life, firefighting, and a twisted sense of humor. We both had a son and a daughter, and a New York born wife to keep us on a straight path. Pete and I would compare techniques used to find and pull people from burning buildings. It was uncanny how many things we both shared interests in both on and off the job. All three are gone now, one from cancer and the other two from heart attacks. This job kills you in so many ways.

I also would like to thank the ladies in my life that supported me along the way, First my beautiful wife who has put up with me for 27 years of marriage, I think she has already won a place in heaven. Next, I am grateful to my mother who raised me and taught me right from wrong. My grandmother was also very special to me and influential upon my life. My daughter Katie and her husband Paul, who recently gave us our first grandchild Natalie, already the apple of her grandfather's eye.

And finally I would like to thank my son Jimmy who I raised to be his own man. He is the best son any father could hope for.

To the hundreds of guys I worked with over the years, I hope you enjoy these stories as much as I have enjoyed writing them. I can still see some of the images in slow motion in my mind; I can still taste the smoke and see the raging flames. Do what you love for a living, and you will never work a day in your life!

To those of you still pulling on boots and climbing on the rigs......... Stay low and safe... I sometimes miss the bells, but the smoke headaches, not so much....Jake

- Forward -

I have known Jake for a bunch of years from my days at Kentland; he has a real gift for putting it on paper like so many of us wish we could do. While the names of the departments are different, having followed a similar Fire Department path makes his story that much more enjoyable. I hope he keeps writing and can't wait for the next one!

Ed Smith

- Chapter 1 -

CAPITOL CITY FIREMAN

The evening meal at 5 Engine was drawing to a close as the men were washing the dishes, Gene-O, the cook, was sitting in a chair reading the newspaper and enduring the nightly playful banter, "Who called the cook a mother fucker?" Bryan Lam called out. The usual reply quickly, "Who called that mother fucker a cook?" Laughter filled the kitchen as all 12 men on duty truly enjoyed working with each other.

It was a cold windy January night outside on the Streets of Richmond, Virginia. And unknown to the men of 5 Engine and Ladder Company 1, an arsonist was busy in a large vacant building just four blocks away. The verbal sparring match was winding down when the truck driver Donny Viar commented that it had been a while since we've had a large fire. "Don't even say that," said Mikey Edmonds, "as you will talk one up." With the dishes finished, the men drifted of to different parts of the one story firehouse to read, watch television, or talk on the phone. Several joined Gene-O at the kitchen table for more verbal fun. The TV in the kitchen was on, but unwatched as the group abused each other in a friendly way that only men who know and respect each other understand.

The vocal alarm speaker with the quick electronic beeps that meant the brothers would be going on a run broke the banter. After the quick series of beeps the dispatcher announced "Engine Companies 6-5-10-33, Trucks 1 & 3, Battalion Chief 3 respond to the corner of Foushee & Broad Streets for smoke coming from a vacant building." Men ran quickly from all areas of the firehouse and took their assigned positions on the apparatus. I was driving the hose wagon that night and Lt. Taller was in command of the company. Mikey, Bryan, and Gene-O were on the back step and Grayson Finner was driving the "Engine," a second pumper that followed the hose wagon everywhere it went. We exited the firehouse quickly since it would be possible to beat Engine Company # 6 to the scene.

When we reached Brook Road, I pushed the wagon as fast as it would go as Lt. Taller's foot pressed hard on the siren button. In my mirror I could see Grayson right on my bumper with the 1976 Mack pumper, the piece of apparatus he would use to connect to a fire hydrant and supply us with plenty of water under the proper pressure. The hook & ladder truck of Ladder Company 41 completed the noisy parade racing up Brook Rd. At the end of Brook Rd I turned the wheel hard to the right into Adams Street while watching out for other cars and pedestrians. A quick left into Broad Street and we were on the scene, and arriving simultaneously was the two-piece engine company of #6. The former Charles Department Store had stood on this lot since 1877, and was now vacant with a light smoke emanating from the top floor. Truck 1 arrived and took a position on the eastern side of the 110' wide brick and joist building and began to raise its 100' aerial platform to the 4th floor. 10 Engine and 3 Truck arrived next and 3 Truck set up its 100' ladder on the western portion of the front wall. Lt. Taller led us down Foushee St. to the alley at the rear of the building as access in the front was blocked with a pedestrian walkway constructed to protect people from the renovation work on the building. At the rear entrance we encountered a chain link fence with a locked gate to protect the property from theft. Lt. Taller ordered Mikey to go back for a pair of bolt cutters but he was already heading that way, as he had anticipated the need for forcing entry.

While we waited Captain Bernard Emerson from 6 Engine arrived with his company and a couple of other firemen from 1 & 3 Truck. Just as the truckmen were about to break the padlock, a uniformed Richmond City Police officer showed up inside the fence. He quickly unlocked the gate and let us into the building. Just inside we saw a beautiful blonde haired woman wearing a knee length red dress and high-heeled shoes. We looked at each other and shrugged our shoulders; she was truly out of place in a construction site at 9:00 p.m., but we had a fire to deal with. We went up the stairs located just inside the rear door and ascended to the third floor. The building was about 100'x 110' and a lazy fire was burning on the exposed ceiling 10' from the front wall. The fire was approximately 15'x15' and didn't seem to be going anywhere. Lt. Taller asked me if I had my rope with me and ordered me to lower it out the front window where Grayson had already deployed a two and a half inch diameter hose line. Capt. Emerson took his company back down to the street and began bringing up an inch and a half line over 3 Truck's aerial ladder. We were working quickly to try to get water onto the small fire before # 6, as this would give us bragging rights until the next fire. Competition with other companies is just a small part of the fun that occurs in every city fire department. This was going to be a very small and easy fire to extinguish. 6 & 10 were both companies of fine firemen that we worked with regularly but there is pride when your company gets first water onto the fire.

I was leaning out the front window untying the rope when we heard a series of "whoff, whoff, whoff!" and the back of my head became very hot, like being on the beach on a sunny day. Bryan said, "Jake, look up!" and when I did I could see flames leaping out of every fourth floor window. Pieces of cornice and window frames were on fire and dropping down onto the men on 3 Truck's aerial as they reversed direction to escape the intense heat of the fire. All the kidding and joking ceased as things suddenly turned dangerous. I looked up at the flames now shooting 15' out of every window above and looked at the empty hose in my hands and just tossed the line back to the street below. There was no way our measly 250 gallons per minute would put a dent in this inferno. Lt. Taller gave the order to evacuate the building and we ran for the staircase at the rear of the building. I stopped at the landing and looked back to see the room we had just evacuated flashover into a sea of orange. Fire was dropping down from everywhere, as I wondered if we would even be able to get out of the building. Running down the steps we met 33 and advised them to evacuate. They were advancing another two and a half and the Lt. from 33 said he wanted to give it a shot. We stopped, looked at Lt. Taller, and he nodded for us to help 33 with their line. When we got back to the third floor landing the entire 3rd and 4th floor was really burning.

Dave Pridge was on the pipe and had to turn his head to avoid the heat while about 7 of us were flaking out the line in anticipation of water that would soon fill the line and we wanted to get it into place before it filled up and became heavy. As we crouched over each other waiting for our water and anticipating the hard push into the fire, a propane cylinder exploded somewhere just above us. I quickly ran my hands over my entire body to see if I was hit with shrapnel as the other firemen did the same thing. Everyone was wide-eyed and deadly serious as shit was really happening fast. The Lt. from 33 was screaming on his portable radio for water when the building shook violently again.

Battalion Chief Griggs arrived in front as the explosions were taking place and he didn't need a portable radio to tell us to "GET OUT OF THE FUCKING BUILDING NOW!" Even though we were way back in the rear of the third floor stairs we could hear his voice clearly. 33's Lt. said, "Ok guys, let's get outta here." We began to descend the stairs bringing the hose with us when another explosion occurred and parts of the roof began to drop down onto the floor above us. Our pace quickened and finally the Lt. said," Just take the fucking pipe off and leave the hose here!" Pridge unscrewed the nozzle as he continued down the steps. Soon we were out the back door and back to Broad Street.

Out on Broad Street the scene was surreal; Buildings 120' across Broad were beginning to catch fire and the plastic store signs melted. The hydraulic hose in the knuckle of 1 Truck's boom was on fire and all the warning lights on the truck were melting. I could see Eddie P and Donnie Viar scrambling to try to get the outriggers up so they could move the truck, as 3 Truck pulled away from the building without a tillerman, the tiller seat on fire. Bryan and I picked up the line I had dropped a couple of minutes ago and began to extinguish the fires on both 1 & 3 Trucks. When we hit the trucks with water steam rose up as the vehicles were very hot. We then shot water above the guys from 1 Truck and let it rain down on them so they could complete their work. The steam rising from the street reminded me of a movie set, but this fire was very real. After the trucks were moved we began to wet the buildings across the street, as their window frames were on fire. More apparatus began to arrive as Chief Griggs struck a total of four alarms.

The first 20 minutes of this job were pure pandemonium. But as soon as the fire burned through the roof, the radiant heat diminished since all of that energy was dissipating into the sky. We could tell that we probably would keep most of the damage to the original building. Companies were operating in and around the fire and thick masonry walls kept the fire from spreading through the entire block. Chief Griggs moved everyone back to a safe distance, since the 60' tall walls were in danger of collapsing. For the next three hours we sprayed the fire from every angle, to contain it to the building of origin. As midnight approached, we took turns leaving our post on the stang nozzle (a large water gun that is capable of flowing 1000 gallons per minute) to go to a truck the Red Cross had set up to distribute coffee. When it was my turn, I went over and the Fire Chief Ronald Lewis was standing in line. He smiled at me and asked if I was getting enough action. "Too much I think," I replied. Waiting in line for the first time that night, I realized I was soaking wet, and freezing. My gloves were frozen and when I took them off Chief Lewis reached into his pocket and gave me his dry gloves. I couldn't believe the Fire Chief would be so nice to a lowly private like me. The combination of dry gloves and hot cup of coffee was wonderful.

Finally around four o'clock in the morning, Chief Griggs gave us the order to take up our hoselines and return to quarters. Other companies would remain on the scene for three days drowning hot spots.

As was our custom at 5 Engine, the next morning when the next shift came in to relieve us, we would travel to someone's house stopping for beer, steaks, and whiskey for Gene-O and blow off some steam, We always got together after a tough fire to relive the tales and experiences of cheating death one more time. We speculated what the blonde was doing with the guard, but anyone who works in the ghetto already knows the reason she was there. "Where did she go to?" asked Mikey. "Who the hell knows," Gene-O replied. It wasn't until Bryan said, "You know there were 14 of us on that 3rd floor when the shit took off, if they (the arsonist) had set up the fire plants on the first or 2nd floors, we would all be dead." That's when it hit me just how close we had come to buying the farm.

By noon everyone was passed out sleeping on sofas or the floor. It was always funny when whoever's wife would come in and find 5 or 6 drunken passed out firemen in her house, usually they would raise hell, and kick us out, but somehow later forgive us, although I don't think they ever really understood the need to blow-off steam in this manner.

- Chapter 2 -

The Early Years

My name is James Jacob Rixner. I was born in Pittsburgh, Pennsylvania. I have wanted to be a fireman for as long as I can remember. My father drove trucks for a living, and my mother stayed home and took care of my brother, sister and me. We lived in a large old house in the Point Breeze neighborhood on McPherson Boulevard. We lived in the area where the Industrial giants such as Henry Clay Frick, George Westinghouse, The Melons, and Andrew Carnegie families lived in the late 1800's and early 1900's. The neighborhood was full of large Victorian style homes, many of which had been converted into two and three family homes. My parents leased the second floor from the owners Mr. and Mrs. Gutillia who lived on the first floor. A young couple lived in the smallest apartment above us in the finished attic.

One of my earliest memories is my dad walking me the 4 blocks up to Engine House 16 at Penn Ave. & Lang St. One of his friends worked there in Squad Company # 8. His name was John Houlihan but we called him Uncle Schmo. We would sit at the long kitchen table with all the firemen, as they drank coffee and talked.

The Engine house was a classic red brick two-story firehouse built in 1885 with a huge hose tower that made the firehouse look like a castle. 16 Engine had a classic 1965 open cab Seagrave pumper that sat on the left side of the floor while Squadron 8 had a small panel van that sat in the right-hand bay. Between the trucks two brass sliding poles disappeared into mysterious holes in the ceilings that appeared to a five year old to be 50' tall. I don't remember ever going upstairs to the bunkroom, so it was always a source of mystery what was up there. Later on as a teenager, I would learn that Engine 16 & Squad 8 were two of the busiest companies in the entire Pittsburgh Bureau of Fire.

The Homewood-Brushton area was in rapid decline and fires were plentiful in the 1960's. Squad 8 answered a larger area than the engine company. And these units were known as "manpower squadrons." They carried an officer, 3 firemen and 4 self-contained breathing masks. Their job was to go in and relieve the first-in engine company and take the hoseline further into the fire building.

Engine companies in Pittsburgh had only a Captain, a driver, and one fireman on the backstep. Coming in to a job they would take a hydrant, leaving a man, proceed to the fire and the Captain would pull a line and try to knock down the fire by himself. Back in those days they didn't have the luxury of having a mask in the cab, so they just took a beating until the squad could get there. I learned many years later that my "Uncle" was one of the toughest firemen on the job. He retired in the late 1980's but I still meet firemen from PBF that knew him or have heard of his reputation.

After Dr. Martin Luther King was assassinated, the area we lived in really got bad. I was 6 years old when we moved to a new house in the suburbs of Plum Borough. It was a big change from living in the city. It was here that my father joined the volunteer fire department. He was a member of a new fire company in Holiday Park. Saturdays were their clean-up days and soon my brother and I were regular participants in the Saturday morning details.

Holiday Park's equipment consisted of a 1939 American LaFrance 500 series pumper and a 1965 Ford 750 Pumper. They also had a 1963 Ford rescue truck with advanced equipment of the day… a porta-power and a cutting torch. Not bad for a 5 year-old fire department. It wasn't long until I knew where every tool was and what it was used for, but it would be a long time before I would take in my first real fire.

During the early 1970's my dad would take fire school classes and bring home training books from these schools and from his many friends in PBF. I would read these books from cover to cover, sometimes at the expense of my schoolwork, which didn't please my mother. I remember teaching myself how to pump a fire engine reading these books and then asking questions during my Saturday visits.

The block we lived on had many boys my age and we always had a baseball game going on in the summertime and football games in the fall and winter. Riding bicycles and trading baseball cards filled our days. It was that time in America when mothers sent their kids to play and told them to be home when the streetlights came on at dusk. Disagreements were handled with fists and usually you were friends again with the boy you had fought with by the next day.

I was an avid baseball fan and my grandmother took me to my first game at Forbes field in Pittsburgh to see my favorite team, The Pirates. My favorite players were Roberto Clemente, Willie Stargell, and Bill Mazeroski. It's a shame that players jump from team to team the way most do today. I can't even imagine any of my heroes playing for anyone else.

The Fire Department had a junior fire program similar to today's explorer scouts but you had to be 16 to join and I couldn't wait. I remember coming home from school one day when I was 13 and my mom was excited and she told me that they dropped the age limit to 13 and I was now eligible to join. My dad got me the application and within a month I was now a junior fireman. We had our own meetings, fundraisers, checkbook, and officer corps. The second year as a member I was elected to Captain of the juniors serving under Allen Bender our Junior Chief. We weren't allowed to ride the trucks to calls but we could ride them during other duties such as parades, fund raisers etc. The next year, Allen turned 18 and became a senior fireman and I was elected Chief at the age of 15.

During my two years as chief we pushed hard to build up money in our checking account. We held raffles giving away television sets, hams and sold donuts door to door on Saturday mornings. I remember how generous the local business people were. We had our raffle tickets printed at the local print shop, and the owner always donated them, he refused to take our money. We got our account up to $2,000.00, which at that time was an incredible amount of money.

Some of my junior guys were starting to burn out from all the extra work so we voted to buy each junior member a pair of real fire boots the proper size to fit our feet. I have always believed in keeping the morale of the members up. The next year the Fire Department bought a new Mack 1500 G.P.M. pumper for $57,000.00. With help from the seniors we decided to buy a new electrical generator and several 500-watt spotlights for the new pumper. I enjoyed belonging there and was learning lots of valuable leadership skills that would serve me for the rest of my life, but there was one problem.

Each day after school I had a Pittsburgh Press newspaper route. I had lots of customers and was earning good money, but almost every afternoon during my route, I could hear the fire siren from Monroeville # 5 just over the hill in the opposite direction from Holiday Park.

Monroeville had a 1967 American LaFrance open cab pumper with a Detroit diesel engine that I could hear going up through the gears from miles away. They were running about 400 calls a year compared to Holiday Park's 60 calls per year. The other problem was I couldn't ride to fires with H.P. until I was 18, but in Monroeville you could ride and do everything except go into burning buildings at 16. It wasn't long until I was riding my bike to # 5 every chance I got.

The members there were younger than at H.P. and I was soon part of the gang. One of the Captains there was a stickler for rules and kept after me to join and since I was already riding on calls for a couple of months he demanded I join or quit coming around. This was June of 1978 and I didn't turn 16 until September. No one had asked me my age and they had invited me to ride so I wasn't going to ruin it by telling them.

To join officially you needed to have working papers, this was a form from the school system to comply with child labor laws. I went with fear and trepidation to the school board office to get my working papers, but they would show my true birth date and I'd be sidelined for the entire summer, so what could I do? When I walked into the office there was a girl working there I knew from school. I knew she had a crush on me so I explained to her my dilemma and talked her into changing my birth date by three months on the paper work. It would cost me, as I had to take her out on a date and keep her happy until September. Looking back it was a foolish thing to do but I couldn't see missing out on the entire summer's action.

When school let out for summer break, I practically moved into the firehouse. Monroeville #5 had a nice bunkroom with about 12 beds and usually had at least 6 people sleeping in. Except for Monroeville Fire Company #4 all other firehouses in the area were home response stations, which meant the firemen would respond from home, this gave us a big jump on the other stations.

During the summer of 1978 we bought a Hurst Jaws of Life Rescue tool.

It seemed that we used that tool a couple times each week all summer long. There were a string of bad fatal wrecks that year. I can remember my first fatal accident. It was about 0230 hours when the red phone rang advising us of a wreck in #6's area on old Route 22. We were out the door with the ambulance, engine, and rescue truck in 45 seconds. While enroute the run was punched out on the radio alerting Companies 6 & 5 of the wreck. Number 6 was a home response station and we were on the scene before their first unit was enroute.

A Chevy Chevette had gone off the road struck a fire hydrant and rolled down an embankment flipping several times, landing upright in a creek. The hydrant had broken the underground pipe and water was shooting 10' into the air, just one mile down the road was the main pumping station for the water system and when the pressure dropped, the pumps kicked in sending water 50' into the air with rocks and dirt with it. I got down the embankment to the car the same time as our assistant Chief, Platt. The rear window was broken already and I climbed through it into the back seat. The driver was lying across the seat with her head in the passenger seat, I reached between the bucket seats to check for a pulse and my fingers went through a large hole into her neck. Platt was outside the car next to the closed passenger side window and I told him she was dead. Her neck snapped by whatever punctured her. We retreated to safety as the water and rock were really coming down on us as the pumps kicked in.

Other HFA's (Horrible Fucking Accidents) that summer included a decapitation when a man in a Lincoln Continental ran under a tractor-trailer truck that was backing out of a service station on Rt. 22. A young man riding in the back seat had his brain squeezed out like a pimple when a Jeep CJ5 rolled over on Rt. 286. A pick-up truck that rolled over a guard rail tearing off half a woman's face, and tossing her body through the rear window into the bed of the truck at 22 & 286. I remember she was with two guys and riding in the middle of the seat. The guys were both drunk and when we got down the embankment to her body, she had already relieved her bowels. I came up the embankment and one dude kept asking me why we weren't helping her? "She's dead friend," was all we could say, but the drunk couldn't process it and kept asking about her.

There was a gang stealing cars from the Monroeville Mall and taking small items such as stereos and the torching them in Hangman's Hollow. We had a fully involved car fire almost every Friday and Saturday night. A lot of trash and brush fires, but that summer was dead for building fires. Our Chief even had a pool for the date of the first fire but months went by until September.

It was a warm Indian summer's day when the alert tones beeped twice "3512 Mountain View Dr. dwelling fire, Companies 5 & 4 are due to respond." Mt. View Drive was on the far end of our district, about a 5-mile ride. Bob, our chief, was in quarters and he could make that old chop-top LaFrance go like hell and soon we were speeding down Rt. 22 hanging on for dear life. Turning into the block you could see the smoke in the sky; finally I had my first working fire. I took the hydrant and was preparing to hook up when #4 came around the corner and put their engine in to the hydrant. After helping 4's driver I came up to see a nice contemporary ranch style house with one room burning in the rear. Batty and Ray already had the preconnect through the front door and soon white steam replaced the black smoke. This was textbook stuff. After helping stretch a back-up line (I couldn't go in) I saw the chief standing by the pump panel, I asked him if he wanted me to watch the pump for him. He couldn't believe I knew how to pump. After asking me several technical questions, he trusted me with the pump, he went off to check on the guys and take command of the fire.

School was back in session and it was always a chore for me to get up in the morning, my mother knew I hated school, but she knew if I quit I couldn't become a paid fireman, and reminded me every time. My dad wasn't to happy about my joining Monroeville # 5 as Holiday Park was his outfit so any mechanical breakdown or misfortune at #5 he would tease me about. I would take any insult personally, and we had quite a rivalry as most teenaged boys and their fathers do. It was always a treat when we took in jobs (fires) together. One afternoon shortly after the Mt. View Dr. fire, I arrived home to find my dad waiting at the front door with an official letter in his hand from the Monroeville government. I think he thought I was in some sort of trouble and made me open it immediately, much to my delight it was a personal letter from Bob Harve our fire chief thanking me for my outstanding work at the fire, he especially commended me for taking over the pump for him and allowing him to perform other tasks. My father was both pleased and proud that I was recognized for good work on my first building fire. I later found out that everyone on the wagon that day received a letter from Harv; he was a classy guy, and a great chief.

Bob Harve was a quiet and unassuming guy, he was always straight forward with us and rarely became angry, even under the most trying circumstances, there were however a couple of times I would test his patience.

The fire was located in the town of Wall, a small railroad community at the southern boarder of Monroeville. I was coming home from work on my motorcycle, waiting at an intersection for the green light when our ladder truck went past en-route to a call. The hospital was a couple of blocks up the street and the alarm came in frequently, but was rarely a fire. Curiosity got me and I dropped in behind the truck and followed to see where they were going. The truck continued south on Route 48 past the hospital and I knew they were going to a fire. Rt. 48 winds it way down a long hill and emerges in the Turtle Creek Valley. About a mile from the bottom you can look across and see the town of Wall. When we reached this point I could see a two-story frame house with heavy black smoke coming from the attic. I parked the bike out of the way and scrambled to find some gear, Tom Dembowsky was driving and I pleaded with him to let me borrow his gear. "No way Brother," he said and I went hunting for some gear from other trucks on scene. Pitcairn # 1 had an equipment truck that carried extra coats and helmets so I borrowed from them.

Arriving back at Truck 5 our asst. chief Platt grabbed me, "Jake, find a mask and get up there and get some water on this fire, these idiots want me to set up the ladder pipe." I looked over the truck but all the air packs were gone. The chief from Wall had all his guys spreading covers over the furniture to protect it from water. There was an inch and a half line lying at the front door so I picked it up. Looking around for someone else from 5 to go with me, I only saw a couple of guys from Pitcairn #1 standing around, so I grabbed the closest one to me and told him to come with me. We reached the second floor and this guy is working his ass off feeding me hose. At the bottom of the steps to the attic was a window so I stopped and opened it. I opened the door to the attic and everything half way up the steps is burning like crazy, I look back at this guy and his eyes are wide as shit. "Are you ready?" I asked him, as he's breathing air through his mask rapidly. He nods his head and I tell him, "Let's go." I open the pipe and the fire darkens down quickly but the smoke goes all the way down to the floor and breathing becomes very difficult. With my face down at the step treads we inched our way about half way up the steps. I'm trying to remember what my father taught me about breathing slow through your mouth so you don't inhale superheated air when I can feel my stomach lurch. I shut the nozzle off and crawl over top of the guy behind me and make it back to the window when the vomit erupts forcefully. After emptying my belly into the side yard and getting a few gulps of air I went back up the steps expecting to find this dude with the pipe finishing the fire. When I reach him he's sitting exactly where he was when I climbed over him. I try to scoot by him but it's too tight and I get him to lie down and crawl right across his back to the pipe. When we reach the attic floor level most of the fire is to my left and I open the nozzle again and bank the straight stream off the underside of the roof. The smoke descends again making it almost impossible to breathe and we retreat down the steps again, this time my fireman friend goes with me. We land in a heap at the bottom of the steps and I've got to go back to the window for more air, my friend is more than a little bit pissed off and leaves me muttering something about, "all you fucking Monroeville guys are crazy."

Spending twice as much time with my head stuck way out the window trying to find pure air I decide to try it again, as I'm crawling up the steps with my face pressed against the floor like a hound dog tracking a scent, my buddy Gary Cole appears behind me like an angel from heaven, he has his mask on and is carrying a 6' hook and asks, "Rixner, what the fuck are you trying to do, kill yourself?" I tell him to make sure the windows at both gables are broken out so I can get some air. He disappears into the smoke and soon I can hear glass breaking, then I hear him coming towards me and past me towards the front of the house, more glass breaks and the smoke begins to lift just enough that I can get off the steps and onto the attic floor. Gary is tearing apart stacks of burning stuff that was important to whoever lives here. As we work together the mucus from my nose is hanging down and dripping on the floor. Gary tells me that Platt found him and told him to get up here with me. As he exposes fire I use the hose to extinguish it.

Suddenly I feel as if I'm drunk, my head is spinning and my legs feel rubbery. I tell Gary I'm going outside for a break. Outside the building Platt greets me with a slap on the back and I see the aerial truck is set up, but the ladder pipe wouldn't be needed tonight. I rested on the front bumper of the truck for about 5 minutes when I decided to go back and check on Gary. At the second floor doorway to the attic, some big dude about 6' 3" is standing with his mask on, breathing air, I try to slip past him but he blocks me and announces that no one can go up there without a mask on. I try to push him to the side but he's got me by at least 150 pounds and 8 inches. About that time Gary yells down the stairs and asks if I'm all right, my friend turns around and looks up the steps and asks Gary what he wants, when he does this I'm looking at his cylinder valve on his S.C.B.A. So being 18 years old and full of piss and vinegar, I reach down and shut his air off. The guy takes another breath and the mask collapses against his face. Screaming at the top of his lungs and running for the steps he's tearing at his face piece. I laugh at the sight of it and rejoin Gary to finish the overhaul job. We finish our work and return to the street I'm taking off my borrowed coat and helmet when I hear this booming voice yelling, "There he is!" The goof ball with the air-pack in a clear building has his Chief by the arm and he's dragging him up the street towards us. A crowd is starting to gather as Platt intercepts them to ask what the problem is. "That motherfucker shut my air off!" shrieked the Goof. My first pal appears on the other side of the chief and whines, "and he pushed me down the stairs!" Platt's doing some fast talking but it looks like a fist fight is about to break out, but the problem is there are 5 guys from Monroeville and about 50 guys from Wall, and Pitcairn. Tom Dembowsky grabs my arm and says, "How did you get here?" I tell him my motorcycle is about a block up the street. He looks at me and says, "If I were you, I'd get the fuck outta here right now." Already feeling like I've had my ass kicked by the fire, I slip behind the ladder truck and up the street. On my way home on the motorcycle, I feel great, the fresh air never tasted so good.

The next day the firehouse is abuzz with talk of how Jake shut this guy's bottle off in the middle of an intense fire attack. People from my own house who weren't at the fire want me suspended, some want me dismissed. I walk in the firehouse and Chief Harve is there, he takes me into his office, which is totally out of character for him and tells me, "I got a call from The Wall Chief, did you shut someone's air bottle off last night?" I tell him yes and he looks like he's going to faint. "Don't tell me that, I've already denied it!" I give him my biggest shit-eating grin and say, "You wouldn't want me to lie to ya, would ya?"

Eventually the mother hens calmed down and I got back to the daily routine of working all day, stop by and spend the evening at the firehouse, go home to sleep and back to work. My girlfriend had to come to the firehouse just to see me. After a few months it all became terribly boring. I had my mind set on being a paid fireman and it seemed like I was slowly wasting my time. The City of Pittsburgh had a residency rule for firefighter candidates and word on the street was they weren't going to be hiring for three or four years due to budget cutbacks. I was working at a tire store, changing tires and doing other small repairs but I wanted something more. The economy in the region was also in big trouble, people in the steel mills with 35 years service were being laid off, and most of those mills would close down with-in the next five years. It was time to make some big changes in my life or I could see myself at forty years old still living with my mother and making minimum wages. I decided that if I didn't have my firefighter job within two years, I was going to join the Marine Corps. The other decision was to move to Prince George's County, Maryland. P.G. is a bedroom community of Washington, D.C. where volunteers face the same heavy work loads of fire that is usually common only in big cities.

- Chapter 3 -

P.G. COUNTY

Monroeville #5 has had a sister station in P.G. called West Lanham Hills Volunteer Fire Co. 28 since 1974 when Dan Poliner from Monroeville # 5, who at the time was working in the Richmond, Va. Fire Bureau and traveling twice a month back to Pittsburgh. He stopped in at 28 one day and met a guy that would change his life any affect many others including me. This man's name was John Sullivan but we called him Sully. Sully invited Dan to stay the night and ride fire trucks with him and they ended up with a couple of fires that night, and so it began that many Monroeville guys would make the 230 mile trip to Station 28 for more action. The relationship between stations continues to this day.

I remember my first trip to P.G., it was the summer of 1979 when Jay Swartzwelder, myself, and a boy we called Chooky got into Jay's Toyota and made the trip to this magical place. I'd heard so many stories about P.G. that I expected to see a building fire before we even got to the engine house. We arrived about one o'clock in the afternoon and Sully had the 1965 American LaFrance open cab tiller truck alongside the fire house and was practicing using the aerial ladder as a crane to bring an injured person off a roof top. The truck was that classic looking all-American fire truck, but it was lime green. I hated that color, but we were visitors and I'd never ridden a tiller truck before and couldn't wait for the first run. We got out of the car and introduced ourselves to the man who appeared to be the boss. This guy with a big bushy mustache, a leather helmet that was scorched, bent, and looked like it was one hundred years old. It was none other than the legend himself, John Sullivan. "Are you the Quakers?" he asked, making a joke at our expense, we told him yes were from Pennsylvania. And he immediately stopped the drill and began to show us the equipment. After asking us many questions about our level of experience he assigned us to exact positions on the apparatus. I remember he gave me "the can" and told me to stay with him.

One thing that impressed me was the organization. Every position on a fire truck had certain duties and responsibilities and if you rode in that position, you were expected to take care of business. Sully explained the difference between being first-due, and second-due and what to expect. He warned us to keep our equipment on the rig and not to be slow because they got out quickly and didn't wait for anyone.

We joined in on the end of the drill and I was afraid Sully would ask me to tie a knot that I didn't know or that I would do something stupid. The man working the turntable controls was obviously the "paid man" because he was the only one in uniform. Jimmy Latham was a 22-year veteran having started out before the creation of the P.G.F.D. He originally worked directly for Company #3 Mt. Rainer V.F.D. and like many of the veterans, was absorbed into the County government when P.G.F.D. was created in 1970. Jimmy was about 6'2" tall and had thinning blonde hair. He had a dry sense of humor and gave you the impression that he didn't like anyone or anything until you got to know him. Once you understood his personality, he was hilarious to be around. The box came in as we were backing the truck into the firehouse. "BEEP, BEEP, BEEP 7302 Landover Rd. Fire reported in the building on Box 33-02 Engine Cos. 33, 30, 9 Truck 28, Squad 22 respond." Men ran for the truck and soon both jump seats had two men in each and Jimmy driving with Sully in the officer's seat. As we made a left out of the firehouse I could see the tiller man at the back of the truck steering the rear wheels of the 52' long truck. I was sitting in the jump seat facing backwards trying like crazy to get all my gear and airpack on while the other three men "in the buckets" were standing up resting one cheek of their ass on the motor cowling. They had their turnout gear on but not worrying about their air bottle and I wondered why. Jimmy had the truck stretched out and we were blowing through intersections pretty fast. Then over the noise of the sirens and air horns, you could hear the radio "Engine 332 on the scene, side 1 smoke showing from the third floor." The voice was a calm, slow, southern accent, without a hint of excitement. As we turned left into Coopers Lane the others began to don their air bottles. I could feel my heart pounding and hoped I would perform my job well. Coming up Landover Rd. the radio squawked, "332 to headquarters, we have a mattress fire, hold 33 return the rest." Red lights went out and the truck slowed to normal speed as we went by the 7300 block we could see the CF Mack pumper of Engine Co. 30 hooked up to a fire hydrant at the entrance to a large Garden apartment complex, the supply line laid-out by Engine 332 already hooked up to the discharge of 30's pump. This was a hard-charging, fast acting Fire Department. At the next intersection we stopped for the light and an open cab C model Mack pumper With Bladensburg on the door pulled up next to us. All the guys in the buckets were standing in the same fashion and I quickly stood up and took my place on the motor box. The driver of Engine 91 revved his motor and yelled over to Sully "Want to race?" When the light went green we raced them to the next corner. I couldn't believe it. This was going to be a great trip.

My first visit ended without any fires and we loaded up our stuff and left Sunday afternoon dead tired. None of us got any rest and the radio in the firehouse talked constantly. My next visit was about three weeks later in early July. When we got to 28 unloaded our gear and got our riding assignments, I asked where Sully is? "That son-of-a -bitch left and went to 33," said a short dude. You could tell something bad had gone down but we didn't want to ask any more questions. Sully it seemed had very little use for fire house politics and had quit 28 several times and joined Kentland Company 33 just over the hill from us.

Kentland had a reputation of being down and dirty tough firemen who never backed off from anything. Most of their men looked like bikers and wore long hair, earrings long before it was cool, and tattoos, anything to look radical. Their firehouse was in Landover, Md. in the middle of an economically depressed area, a real ghetto. I learned that the short dude was the Captain, Billy Whetzel. Later that night after eating some fast food, another 33 box came out for Virginia Ave. Back onto the truck for the trip that would become so familiar, down Annapolis to Coopers, Left into Coopers lane right on Old Landover, left on Kilmer, left on Route # 202 and into the ghetto. Coming down the hill on Coopers Lane we could see a column of smoke darker than the night sky. By the time we pulled up, only steam was coming from 4 windows that you could see by the scorch marks on the bricks above the 2nd floor windows, heavy fire was showing only a minute ago. We were 2nd due so the 3rd floor was our responsibility. We threw a 35' ladder and went up and searched the entire 3rd floor, 4 different apartments. The search was negative, meaning we didn't find anyone overcome by smoke but the heat was tremendous. After searching, Captain Whetzel led us back to our truck where it seemed half of the equipment we carried was off the truck being used. 5 of the 13 ground ladders were raised against various windows, I would later learn that the driver and the tiller man would throw these ladders to rescue anyone trapped and to give us, the searching firemen a second way out in case we got into trouble. Both preconnected cord reels were stretched into the fire apartment providing lights for the firemen in there. Fans were in windows exhausting smoke. People think a ladder truck is used mostly for the main 100' aerial ladder, but I would soon learn that it is more like a rolling toolbox. Lights, ladders, fans, and hand tools get used much more frequently.

Directly in front of the fire building is a white Seagrave pumper with an 8" horizontal orange stripe. This fire truck looks like it has been in a demolition derby. The front bumper is crushed, every piece of glass has the spider web looking cracks, the seat covers were torn and foam rubber was missing from most of them. Both cab doors had swung way past their normal stopping point and the mirrors on the doors rested on the windshield. Standing next to the pump panel is my buddy Sully; he's wearing his normal navy blue uniform shirt and pants, his seasoned leather helmet which now has a bright new "33" centerpiece that looks out of place on the baked and burnt lid. He has a big shit-eating grin. "What's wrong Quaker, you boys at 28 a little too slow to see any fire?" I walk over to him shake his hand and we talk for a while, he invites me to come down to 33 and spend the night, but I don't feel right leaving 28 on this trip. I promise to make my next trip to Kentland. While we are talking I notice none of the gauges on the pump panel are working. "Sully, what's wrong with the gauges?" I ask, and he laughs loudly and says "you're not in Quakertown anymore kid, this is the ghetto, we don't buy fire trucks here to look pretty; here we run them till they die."

"But how do you know if you're giving the attack lines enough pressure?" I ask. He laughs again and says, "I just rev it up a little 'til it sounds right, they will holler if they need more water." Looking back at that night I realize how green I was, a good fireman learns how to adapt, improvise, and overcome any obstacle.

Each Trip to P.G. made me want to leave Pittsburgh more and more. Although it is a great city, and both the Pirates and the Steelers were winning championships, I couldn't find a good job. So finally in November of 1980 I made my decision to move into fire Station 28; going from a firehouse doing 400 runs per to one doing 1800 runs per year. They let you live in the firehouse rent free, but in return you're expected to provide staffing for the fire trucks and hold a steady job. To an 18-year-old kid, this was like having your cake and eating it too. My girlfriend was going to college at Clarion State University about 75 miles away so that wasn't a factor. The hardest part would be leaving my friends and family behind. I set a date for my move; January 1, 1981, New Year's Day, seemed like a good day to start off on a new course, one that would lead to a firefighting career or the U.S. Marine Corps.

- Chapter 4 -

THE BIG MOVE

Early Thursday morning, January 1, 1981, I set off to a new beginning. I was driving my mother's car. I had an army style footlocker with all my clothes in it and a grand total of $400.00 in my pocket. I arrived at fire station 28 at 1300 hours (1 P.M. for you civilians.) Billy Whetzel assigned me a bed in a cubicle on the second floor of the fire house. I was assigned the layout position which meant I rode the backstep of the Hose Wagon and would toss the 50' lay-out section into the street when commanded to do so.

I got busy unloading my stuff and was uninterrupted by any alarms. At 2139 hours the alarm sounded for a first due box. 5319 85th Ave. for smoke in the apartment building. We were out the door fast and I was hanging onto the crossbar on the backstep of the wagon. 85th Ave was famous in the early 1980's for large working fires. Water supply was also a problem as the mains in the 5300 & 5400 blocks were far too small for the massive wood frame apartment buildings.

As we arrived in the block, the wagon slowed quickly and the officer yelled from the front seat to, "LAY-OUT!" I grabbed the rope strapped to the end of the 3" hose line used to supply the pumper with water and tossed it into the street. The fire truck never stopped rolling during the process and we were soon in front of a 3 story garden style apartment building with nothing showing. The officer yelled, "250'!" as he got out of the front seat. I pulled the 100' of inch and a half diameter hose and placed it on my shoulder, stepping down from the rear step; I reached up and dumped the remaining 150' of hose in the street. Then I proceeded toward the building holding the 100' of hose on my shoulder tight, until the 150' of hose was stretched out on the ground. As the firefighter approaches the apartment on fire, he needs to estimate how much hose is needed and about 75' from the fire the firefighter starts to deploy the hose from his shoulder. If he doesn't, he will end up with too much hose and a pile of hose and no water. At the front door of the building Brian tells me to, "HOLD UP!" We wait while the officer investigates the odor of smoke on the second floor. Other fire trucks arrive and take their assigned positions and soon we hear the report of a "furnace motor hold 28; return the rest." The other fire trucks pull off as we start to pick up the 700' of hose we have put in the street.

I was now a live-in fireman, and the runs came in often. The first month I lived in the fire house we had 121 calls and 9 working fires. I made 61 runs and 7 working fires. Some firemen don't see this much work in a year. I was having my cake and eating it to. Each day was a new adventure and I was learning more and more on each call. But aside from some trash in the basement at 5802 Annapolis Rd. and a couple of box alarms a day in Kentland, it wasn't until January 9th that we had a working fire. At 0305 hours the radio beeped three times and announced "7600 Fountain Bleu Dr. Smoke on the second floor on box 28-12, Engine companies 28-30-13 trucks 7 & 1 respond." We were out the door in less than 45 seconds with Dickie Dunlap our paid man driving, Sully in the officer's chair, (he was back from 33) and me on the backstep. Turning into Fountain Bleu Dr., I was ready when the order to lay-out came and dropped the hose directly in front of the fire hydrant that Engine Company 30 would soon hook-up to and use to supply us with water. 7600 was a high-rise apartment house and we grabbed the high-rise bag that contained 150' of inch and a half hose we would use from the standpipe in the stairwells. Reaching the second floor we could smell the unforgettable smell of burnt food. It is a rank odor that every fireman soon learns very well. We force the door to apartment 2-b and find the occupant passed out on the sofa with a pot of unidentifiable meat burning on the stove. We have to yell very loud several times to awaken the drunken man. When he awakes he takes a defensive posture as if we are about to rob him. I go to the windows and begin to open them to allow the smoke to clear. Sully radios the chief and tells him we can handle it with the first due truck. Truck 7 comes up with an electric fan that moves 500 cubic feet of air per minute and soon the air is clear in the apartment. The drunk having had time to regroup thanks us over and over for saving his life. As we are putting the standpipe bag back on the wagon we hear "Truck 1 to Headquarters

Priority." "Go ahead Truck 1," the dispatcher responds. "Finns Lane & Riverdale Rd. on the scene with a one story house well off!!" We break the connection of the 3" hose and leave all 600' of it laying in the street as we speed off for Finns Lane. Truck 7 is directly behind us in their open cab CF Mack ladder truck. Pulling up on the scene, fire is shooting out every window in the house. Engine Company 13 already has a line through the front door. We stretch our line to the front door to back them up. As soon as we reach the front door, 13 is backing out and the officer advises Sully that it's too hot and there are holes in the floor. Sully curses the man and shoves his way past him and yells "Quaker, get your white ass up here". I scramble to the front door and we both stop to put our face pieces on. As we are doing this, Dickie charges our line with water. We enter the house together and about 5 feet in, my leg goes through the floor. I can hear Sully cursing as I bring my leg out of the hole, and through the hole I can see the basement is well involved in fire and tell Sully. He tells me, "Yeah, I see it, now let's go get this motherfucker!!!" It is so hot that I feel as if we are both going to melt into our boots, but Sully keeps urging me forward. "That's it! Keep it moving," he says. We continued to crawl across the living room with each of us going through the floor about every three steps; I am sweeping the stream of water back and forth across the ceiling as we inch our way forward. Down the hallway and stick the nozzle into each of the three bedrooms and bathroom. The fire dies down quickly and soon we are at the basement door and Sully takes the pipe from me and tells me to feed him hose and dives down the staircase. I have trouble keeping the hose moving and can't believe what I just saw; it was the gutsiest thing I'd ever seen. I try to get down the stairs but the steam drives me back twice. On the third try I hit the basement floor and follow the line until I reach Sully, His face piece is off and he's grinning ear to ear. Two floors of fire are now knocked down and Sully is as happy as I've ever seen him. "Quaker, you did a fine job." Man that felt great to hear. "Now let's go out and take a break while those pussies from 13 clean up our mess." I had just been taken to the John Sullivan School of firefighting, and passed the test.

A rash of fires started in the Brightseat Road area on Saturday night January 10, 1981. Brightseat Rd. was packed with run down garden style apartments. The buildings were arranged in groups with cul-de-sac streets serving them. It was a perfect set up for selling drugs and other criminal activities. The buildings were occupied by poor people, and were packed with children.

The first box alarm of the night came in at 2103 hours for a fire in the building at 8427 Hamlin St. We made the trip on the truck and arrived to see the Kentland firemen had already knocked the fire down, we overhauled the terrace level apartment. Leaving the fire ground, we were flagged down by a citizen. A fire was burning in the rear bedroom on the first floor at 3034 Brightseat Rd. A quick search was conducted and when 33 arrived the fire was handled. Kentland was always a great group of men to work with. They didn't accept excuses. Either you put the fire out, or got your ass whipped. It was very simple, one thing you could always count on was the boys from 33.

A feud was brewing for quite some time between Companies 33 Kentland, and 38 Chapel Oaks. On January 19th it boiled over when the 33 box was struck for 2262 Brightseat Rd. When we arrived on side 3 for the terrace level apartment, fire was burning in a couple of rooms. 33 had already gone on the scene side 1 nothing visible, and 38 quickly followed us into the alley and placed a hoseline in service through the open sliding glass door. As soon as the fire was knocked down we began to throw the still smoldering mattresses out the window then some words were exchanged between a guy from 33 and another guy from 38. The dude from 38 slowly took off his backpack style air bottle, turnout coat, and sucker punched the guy from 33. All hell broke loose; men were punching each other and paired up one on one in fisticuffs. The fight spilled out into the rear yard and the crowd that had gathered to watch the fire was now treated to a fistfight. After about two minutes of free for all fighting Jeff Pitts from 33 stood up wiped the blood from his mouth and looked at the firemen from 28, and 8 and said " All you guys from 33 & 46, and any of you others that want to help, get some tools and were going to kill these motherfuckers." A group ran over to our ladder truck and stripped all visible tools and soon the two groups were going at it again, this time with pike poles and axes. I saw one guy get hit in the head with the side of an axe; he fell to the grass unconscious. This was some unbelievable shit. Our Captain Mark Larry was trying to round the guys from 28 up and push us back into the building, but there was no way we were going to miss this fight, when out of nowhere he is punched in the mouth knocking three teeth loose. He spits out his teeth and soon we too are part of the brawl. Somebody gets on the radio and puts in a mayday and says "HELP! GET THE POLICE HERE, KENTLAND IS TRYING TO KILL CHAPEL OAKS" within seconds sirens fill the air and police cars swarm in from all directions. The back yard looks like a war zone men are lying on the grass bleeding, two are unconscious, those still standing are holding various body parts which are bleeding. The first cop that pulls up asks, "What's going on?" Jimmy Wells from 33 replies, "Nothing," as he wipes his bleeding mouth. It was the most comical understatement I'd ever heard. Ambulances carry the wounded away and the police let everyone go with a stern warning.

The next day in The Washington Post there is a story about the fight along with a cartoon showing a house burning in the background while the firemen fight each other. I knew when I moved to P.G. life was going to get interesting, but I was wondering if I had bitten off too much.

Things settled down and I slowly became used to the constant radio chatter and was able to sleep with one eye open. To satisfy West Lanham Hills Fire Department requirement for live-ins I had to find a job with-in 30 days. This was not a problem as I had limited finances and soon found a job with the Hunterman's Ambulance Service on Georgia Ave. in Washington, D.C.

Hunterman's was a family owned business that mostly transported people from hospital to hospital for tests, and people from hospital to home or nursing facility. Pete was the alpha male there and his sons Mike & John each drove custom made ambulances with all kinds of bells and whistles. Sort of like the 1970's movie "Mother, Jugs & Speed". Toni was the divorced wife who came in each day to do the books, and was just as sweet as she could be but the constant bickering that went on was incredible.

I was working with John in his customized ambulance on March 30, 1981. We were pulling into George Washington Hospital when the radio put out a call for a triple shooting at the Hilton Hotel not very far away, in Fact we had just drove past the Hilton minutes ago. Once inside G.W., we took our patient to the C.T. scanner room on an upper floor and as we were coming back out we were rerouted by a secret service agent with an automatic rifle. Ronald Reagan had been shot at the Hilton and was brought in by Presidential limo and the hospital was locked down. We were sent out through a side door and had to roll our stretcher down the sidewalk around the hospital. On the building across the street we could see a sniper had set up on the roof. John made the comment "Man lets get the fuck out of here." I was in full agreement. We went over to D.C.F.D. 23 Engine house and were telling the guys there about President Reagan. "No he's ok it's on T.V." "No he's not; he's in the operating room right now." John told them. I called my dad at home. He had been off sick, having been diagnosed with colon cancer, and was watching T.V. at home. "Dad, President Reagan was shot." I told him about how we were 5 blocks away when the call came in and how we were redirected in the hospital. "No, no," he said, "the president is safe back at the White House." "No Dad, he's on the operating table as we speak." My father couldn't believe it but I could understand that during the cold war, information must be protected until they found out who the shooter is working for. This was my first lesson of many on not trusting what you hear on television. It was about 15 minutes later when the press released the truth about the shooting to the public.

As I settled in to firehouse living I began to apply to the paid fire departments in the D.C. metro area including Anne Arundel County, Howard, Fairfax, D.C., Baltimore City. During this time my buddy Dan Poliner talked me into driving 2 hours south down to Richmond, VA. He advised me that they have a great fire department. So In early spring I set out for the City of Richmond.

I arrived in Richmond at 10 am. Getting off Interstate 95 on the Broad St. Exit I drove west through the city until finding the Firehouse at Lombardy & Broad Sts. 10 Engine and 3 Truck shared the gothic two story firehouse built in 1906. I walked through the open bay doors and asked the first man I saw "Where do you get an application?" Charlie Bennett walked me outside and pointed east to a white high-rise building 25 blocks away, "See that building?" "Yeah," I replied. "That's where you need to go". (Years later Charlie would break me in as a new Lieutenant working with him on the same shift at 1 Engine, and I at 2 Truck.)

The firehouse was the classic turn of the century building, when cities built edifices of character and distinction, arched brick bay doors and second floor widows hinted to a time when men went to great lengths to build things right.. The brass poles were polished and everything seemed orderly and clean. All the men had uniforms and badges on; this was something I had never seen before.

The Engine Company had a Maxim hose wagon and an Oren pumper, the truck was an American LaFrance tractor drawn aerial similar to 28 Truck in P.G., but this one had a roof, and was the proper color. I thanked him and decided to walk the 25 blocks to get a feel for the flavor of the city. I could see plenty of potential in the buildings I passed. Most were built in the late 1800's and at least one was gutted by fire every two blocks. Yeah, these guys saw some decent work.

When I arrived at City Hall sirens filled the air as I climbed the steps. 3 Engine and 1 Truck sped down Marshall St as 6 Engine and Tactical Squad 1 screamed up Broad St. A few minutes later 5 Engine sped by. I would later learn that Medical College of Virginia Hospital's alarm had gone off 3 blocks away bringing 5 engine companies 2 trucks and a Battalion Chief. This was a daily run for several years. But at that moment it was an omen for me. Seeing that 1953 Peter Pirsch Hook and Ladder of 1 Truck speed by with the tiller man sitting inside the aerial ladder. The two piece engines of 3, 6and 5, all classic American fire trucks. I knew right then, I was meant to be a Richmond Fireman.

I went inside and filled out the application. Leaving City Hall, I headed west on Marshall St. At 6th St. there was an old armory building that took up half the block. A fish market occupied the corner space and the aroma of fried fish filled the air. I entered the market and ordered a sandwich. While waiting for my order the long red hook & ladder wagon of 1 Truck passed by going up 6th St. I ducked out the door to see where they were headed when their red warning lights came on and they stopped at the opposite end of the building. They got off the truck and the truck backed into the Armory building, the tiller man leading the way. The Korean owner of the market was calling me back into the store and gave me my order. I took the paper bag containing the fish sandwich and walked down the sidewalk to the quarters of 3 Engine and 1 Truck. The firehouse blended in with the neighborhood so well, I'd have never known it was there if 1 Truck had not have come back from their run. I looked around for a few minutes and since the troops were all in the kitchen eating their lunch I headed back towards my car. And back to P.G. County.

Prince George's County, Maryland, has a unique dispatching system. Loud speakers blare through-out every engine house in the county 24/7 and you soon learn to sleep with one eye open. For a fire assignment, two quick beeps followed by the address and nature of the fire and then the 2 engines and 1 truck companies that are due on the call. A box assignment is the same only it has three beeps and gets more units going (3& 2) When you hear the beeps followed by a hundred block similar to those in your area, you instinctively sit up in your chair and prepare to run. It is sort of like a horse in the starting gate at a big race. John Sullivan, Steve Sybolt, & I were sitting in the day room at 28 one afternoon when three beeps came out of the speaker, 5910 so and so street, fire in the building on box 29-31 We all had started to take off but it was a Southside box far from our area. Sully laughed and said "I though it was going to be 5910 Princess Garden Parkway." We all laughed because everyone was thinking the same thing. 5910 Princess Garden Pkwy. was a 6-story hotel and we ran it from time to time for alarms and such. He no sooner got that out of his mouth when "Beep, beep, beep, 5910 Princess Garden Pkwy. Fire on the 6th floor on box 28-12 Engine Cos. 28-48-30 Trucks 18 & 7 Squad 14 respond..." Now back in those days the squad being added to the box by the dispatcher was a signal that they were getting lots of calls and we probably would be going to work. I was "running the line," meaning I had the nozzle and was in the bucket seat on the wagon as we sped down Annapolis Rd.

Arriving with nothing visible from the street, I grabbed the standpipe bag. The bag contained 150' of inch and a half hose, a gated wye, a short section of two and a half, and some wrenches. I was running with the bag on my shoulder trying to keep up with Sully and we climbed the steps quickly, a little too quickly and when we reached the 6th floor I was out of wind. Sully opened the door to the hallway and thick black smoke poured into the stairwell. I was gasping for air as we opened the bag and I gave the pipe to Sully and I figured I could stay here and make the connections while I caught my breath. Sully disappeared into the black wall of smoke as I fed the line to him. He was calling for me to turn on the water before I could make the connection, and I had to hustle up, since I didn't know how bad it was down the hall. After finishing and turning on the water, I donned my face piece and started down the hall towards Sully. I was groping along when I tripped over something lying in the hall. Reaching down I felt the body of a man, when I touched him he moaned and said between coughing fits, "I say old chap, I'm having a bit of trouble breathing." He was obviously an Englishman and was barely conscious. I took off my mask and placed it over his face. Then with him wearing my mask, I began to drag him down the hallway. He seemed to weigh a ton and the carpet on the floor was increasing the friction. I was hacking and coughing and when I stopped for a moment to retrieve my mask, the Englishman had a death grip on it and wouldn't let it go. I continued until I got him to the steps where he promptly jumped up and ran down the stairs like a scared rabbit. I re-donned my mask and went back down the hall for a second time, this time I made it to Sully who I found laying against the wall of the burnt out room with that familiar shit eating grin. He had knocked the fire down and I told him about my grab. He radioed down for the medics to look for the guy, as he probably needed medical attention. They never found the guy, and boy did I catch hell from the other guys for my "ghost rescue." But it felt great to make that first grab.

- Chapter 5 -

Testing for a Job

During the next 15 months I would take a day off from working on the ambulance to test for paid firemen jobs. Pete Hunterman was very understanding and instinctively knew I wanted to make something of myself. I tested for Anne Arundel County, Md. and came in 3rd; they hired two from that list. Howard County's test resulted in nothing; Fairfax, Va. looked promising as they were projecting to go over 1,000 paid men during the next budget year. Richmond called and the written test was scheduled for December 1, 1981 at John Marshall High School. I was served with notice that I needed to report for an eye exam at city hall at 0800 hours. The written test was at 1500 hours. It was a dull, dreary, rainy day. I took the eye exam and passed with flying colors. Now what should, I do with the 6 hours until the written test. I drove all around my new city, in my mind I was already a city fireman here, it just felt that right. It's like the first time I saw my wife, love at first sight. Same thing here, when something is right, you know it in your heart.

I ended up at 1 Engine and 2 Truck at 24th and Broad. The Engine was out on an apartment building explosion in Mosby court housing projects. And the truck was home alone. (Captain Baker from 1 would win a medal for a rescue at this fire) A guy from 2 Truck named Dean Hazlewood showed me around, and when he found out I was in town to take the test he invited me to stay for lunch, one catch, I had to go get the groceries. Dean rode with me down to Grubb's market in Fulton Bottom. We bought the food and returned to 2 Truck where he cooked

31

lunch. After the meal I thanked the guys and rode by the explosion on 19th Street. I visited 5 Engine house on Leigh St. but only 6 Truck was there as 5 was at the explosion, and then went to 14 Engine at Chamberlayne and North Avenues. 14 was in a house originally built for Henrico County and annexed into the city. It was a small 2-story house with one brass pole. Captain Hall was in command of the company and made a big deal about a visitor from out of state. After being treated like a VIP, I reported to John Marshall High at 1430 Hours.

The line waiting to get into the test was about three city blocks long. I couldn't believe the amount of people wanting to take the test. Finally at 1500 we were let into the school cafeteria, given instructions and the test began. I was due in to work at 1800 hours at the ambulance service and knew it was a two-hour drive. Like a fool I rushed through the test, being the 2nd person to finish. On the drive north to D.C., I was kicking myself in the butt for rushing through the test. Here was the job I'd dreamed of and I let something like a minimum wage bullshit job mess it up! I was so sure that I'd screwed up the test.

A week later I received a notice to report for a physical agility test a couple of days before Christmas. I was sitting in the day room at 28 when Bill Shoeman asked, "When is your physical agility test?" "Tomorrow at 0900," I replied. "You stupid motherfucker, you're going to sleep here, take the chance of catching a fire and being tired for the test." I knew instinctively he was correct and packed my stuff. I drove down to Richmond and with my limited funds I stayed in a hotel room near the airport on the east side of town.

The morning of the test I reported to the fire-training academy and about 6 people were reporting every 15 minutes to take the test. It consisted of things like walking the beams of a ladder turning around and returning without letting your feet touch the ground; walking up 4 stories with an air-pack and pulling up a section of dry two and a half inch hose line and running for 12 minutes. I completed the test and drove for six hours to Pittsburgh to spend time with my family and friends.

At Monroeville # 5's annual Christmas Party everyone wanted to know how things were going. It was good to see old friends and I was on the fast track to making something happen in my life. More friends had been laid off and news wasn't good for the region. Monroeville and the entire Pittsburgh area depended on the steel industry, which was slowly going belly-up. In a way it was depressing, but I felt fortunate, at least I was young and didn't have kids and mortgage payments like so many of my friends. In short nothing was holding me back.

After a couple of weeks I received letter from the city, it seems I scored well on the fireman's test and was ranked 5th out of 1200 applicants. Maybe rushing thru the test wasn't a bad idea. In April I received another letter directing me to report for an interview at the fire chief's office. During the same time I was completing the hiring process in Fairfax County.

On a bright and sunny day I reported for my interview in Richmond. Upon entering the Fire Chief's office I identified myself to the secretary behind the desk. Nancy Hale had worked here since 1966 and was the defacto person who decided if you got hired. She had a lot of power and even had a fire helmet with the I.D. of 50 and a half. 50 was the radio signal for the Fire Chief. Nancy asked many tactful questions as I waited, and I quickly realized that this was the real interview. Soon I was called into the conference room and two captains, one black and one white asked me a series of questions. They played the classic good cop, bad cop routine and the white guy was the bad cop. One of the questions was on how to search a room, the white captain ordered me to get down on my hands and knees and show him how to search a room. I was wearing my only suit and wasn't about to go crawling around on my hands and knees for anybody. I told him so and described verbally how to make a search. The interview ended and I wasn't too sure on how I'd done.

Fairfax County's process was wrapping up, and my interview there went much smoother. It was a panel of men from different ranks; they asked me questions about my background. I answered them easily since I had previously filled out a 7-page personal history form that the Fairfax County Police detectives had checked out. There wasn't any sense lying to them, as they already knew the correct answers. When they asked if I drank, I told them only occasionally with the boys, they all laughed. I was sure they knew of the bar room brawls I'd been in.

In early May I made a trip to Pennsylvania to attend the wedding of Bruce and Debbie Hensler. When I got home there was a letter from Richmond that my mom had opened. It said that although I was well qualified, no job was offered to me "at this time" I was upset that I didn't get the job. I knew from talking to Nancy, they were hiring 9 firemen. I was 5th on the list; I couldn't understand why I'd been skipped. I sat down right then and typed a letter to Chief Lewis in Richmond, VA. I was concerned that I was doing something wrong and didn't want to make the same mistake in Fairfax. I asked him what I'd done wrong and asked that he be frank and brutally honest with me. Two days later I was at the firehouse in Monroeville when my sister called, "You need to call Chief Lewis in Richmond," she gave me the number and I quickly dialed it. Chief Lewis got on the phone and asked me, "Mr. Rixner, do you still want to work for me?" I couldn't believe my ears, "Yes sir," I replied. I sent you a letter two days ago; if you haven't received it yet please disregard it. I went to Bruce and Debbie's wedding the next day, but I can't honestly say I was celebrating their wedding.

I started recruit school on June 14, 1982 a week later than the other students. About two weeks into school I got a letter from Fairfax County. I had been appointed to the Fairfax County Fire Department; starting July 26, 1982 this was too much to believe. I had looked for work for two years, and now within two weeks I had two fireman's jobs

In my scrapbook at home I have two letters, one from Richmond saying I didn't get the job, the other from Fairfax saying I did. An old friend, Lynn McConahey was working at Vienna Station 2 in Fairfax and was begging me to come to work for the county. I asked him, "Exactly what do you do?" "One day I drive the medic unit, next day I am in the ambulance, the third day I am on the hose wagon." "So one third of the time you're on a fire truck." All around Richmond I could see burned out buildings, "No, I think I'll stay here," I told him.

In their letter they had a deadline for accepting the job. I waited until the last possible day and sent them a no thank you letter. I never said anything to anyone in Richmond about the Fairfax Job so I was surprised when on the Monday morning of the 26th our drill instructor, Lt. Beasley meet me at the entrance door to drill school with a big grin on his face. "We didn't expect to see you here this morning," he said. "Why?" I asked? "Fairfax called us a month ago; my money was on you going to work for them!" "No Lou, I think the fire duty is better here." "Where do you think you want to go?" he asked me. "I'd like to go to a busy engine company," I replied. " I expect you will do alright," he said. Once again everything felt right.

Drill School was held Monday through Friday from 8am to 5 pm. and was 12 weeks long. There were 9 men in my recruit class and we were taught the Richmond Fire Bureau way of putting out fires. I knew from experience that all fire departments had their own way of doing things, and no one wanted to hear how Prince George's County, Maryland operated. When people in the class asked about my previous experience, I brushed it off. The drill instructor seemed to always be within earshot and I quickly realized this pleased him.

As the school progressed I got the feeling that I was becoming the favorite student of the drill instructor. Anytime there was a challenging task, Lt. Beasley would ask me to try it first. For example, the hook & ladder trucks all carried a large round net for people to jump into from upper floors. It was very old technology and most cities didn't carry them anymore as it took 7 to 9 men to hold the net, and there was a real danger of being hit by the person jumping. The class practiced jumping all day long from the 3rd floor windows. Lt. Beasley asked me if I'd give it a try from the 4th floor. Although the 3rd seemed pretty high, it was the way he asked me the question that I knew I had to try. I ran up to the 4th floor and when I looked down at my classmates holding the net, my heart skipped a beat. For a second I didn't know if I could do it, but I figured that if Lt. Beasley had asked me, then I must be safe. I jumped and knew I was going to be alright before I hit the net. Class continued and we learned all the things a rookie fireman had to know about water pressure in hose lines, raising ladders, wearing SCBA (Self Contained Breathing Apparatus), and other tools and appliances.

The last week of drill school we had a final exam; Robert Crass, Chris Garnell and I were within tenths of a point from each other for valedictorian. I happened to stop by 10 Engine to visit the night before the test. Captain Bernard Emerson was at the kitchen table going over the day's memos ands one of them dealt with the Fire Chief's radio I.D. being changed from Unit 50 to Unit 100. The next day at drill school one of the 100 questions on the final exam was what's the fire chief's radio number? I got it correct and it was the difference between 1st and 3rd place. I can't say what made me stop by # 10 that night but it made all the difference in the world.

During the last week of drill school Lt. Beasley read the transfer list from the fire chief. It listed our new assignments starting next Monday morning. As he went down the list calling each name, moans or shouts went out from each man depending on whether he liked his assignment. Finally, I was the last on the list, Lt. Beasley called my name and paused, the suspense was killing me. Probationary Fireman Rixner... Engine Company 5 "B" platoon. I was grinning from ear to ear. Engine 5 was the busiest hose outfit in the city.

After we were dismissed for the day I asked to speak to Lt. Beasley for a minute, I thanked him for helping me get assigned to 5 Engine. He told me he was assigned there when first promoted, he filled me in on the type of guys I'd be working with, and how he was friends with the captain there. Then he told me something I'll never forget. "You're going there on my strong recommendation, I don't want to hear anything but good things about you from Leigh St." Then he smiled and said, "I'm sure you'll do fine."

Friday August 27, 1982 was a beautiful bright sunny day. My girlfriend, mother, grandparents, and other family members were in town to attend my graduation from recruit school. The past 11 weeks consisted of getting up in the morning, going to school, returning to my apartment, fixing something to eat, studying, and going to sleep on the mattress on the floor I'd borrowed from 28. Sully had lent me a 6" black

& white television set to use, but aside from that I didn't have a stick of furniture. I'd been too busy to even care but when my grandfather saw the empty apartment he grumbled about how I'd been living. "You don't even have a table to eat on," he said. The ceremony was held at the Virginia War Memorial on Belvidere Street. The Mayor of Richmond made some remarks about the bravery of firemen and how we were now to go out and serve all the citizens, both rich and poor, He commented on how this class had the highest test scores ever in the history of the Richmond Fire Bureau. Then he called my name and told me to come forward and presented me with my Valedictorian award. At that moment I was the happiest man on the planet. Then they awarded the 2nd place to recruit Christopher Garnell. Afterward, each member of my class was called up and we received our diplomas. We were now "City Firemen".

Chief Ronald Lewis gave an inspiring speech about how we were only beginning a long journey, and how we were to strive to improve ourselves during our long careers. He advised our families of the stress inherent with the position, and urged them to support us. The sun beamed down on us as the entire class with our training academy staff posed for pictures in front of the Virginia War memorial's walls. When we returned to my apartment, there was a new dining room table and a set of chairs, a gift from my grandfather.

The weekend went by too fast and on Sunday afternoon my family was loading up their cars for the trip back to Pittsburgh. My grandmother asked, "When do you start work?" I explained, "My shift is working today and that we start a four-day break tomorrow morning. That means I don't report until Friday." She was smiling and said "Geesh, your first day in the fire house and you're off all week; what kind of easy job is this?"

- Chapter 6 -

First Day in the House

Friday morning I awoke at 0530 hours without the aid of my alarm clock. I was so pumped up and couldn't wait to get in to the firehouse. My apartment was about 30 blocks away and the drive took less than 5 minutes. Parking my car in the alley behind the firehouse, I entered the back door to find about 7 men sitting around the kitchen table drinking coffee.

Right away the teasing started. "THE ROOKIE IS HERE!" one of them shouted as I walked over to introduce myself. After meeting all of the guys at the table, they told me to go up front to the office and see the lieutenant. I went out to the apparatus floor and passed another guy on my way to the front. I knocked on the office door and was told to come in. Sitting at his desk was my first boss Lt. Al Mosley. He rose from his chair and extended his hand. Lt. Mosley was a black man who appeared to be about 30 years old. He asked if I had my equipment on the apparatus yet. "No, it's still in my car," I replied. "Well go get it and put it on the wagon," he told me. I went to my car and brought in my thigh high boots or ¾ boots as we referred to them, my turnout coat, and helmet liner.

Lt. Mosley was waiting for me by the wagon and gave me a helmet shell with blue # 5s on each side. I assembled my liner and shell together and placed my gear on the wagon. The "loo" then took me back to the office where two officers from 6 Truck were exchanging information. That was the normal procedure for the off going officer to brief the in coming one on anything of importance, department information, apparatus issues, fires, etc. The two lieutenants shook my hand and welcomed me to the Bureau. Lt. Naprokoski was in his late 50's and asked me if it was true that I was from Pittsburgh? "Yes," I told him and he told me he grew up there too, leaving to go into the service and then he met a girl from Richmond and after the war they settled here. It made me feel instantly at ease. While he was talking I was doing the math in my head, if I had guessed his age correctly, he was probably a World War II veteran. The other Lt. had gray hair and I guessed him to be in his mid forties, and he had a slight smile that I instinctively knew meant he liked to stir trouble. It was the type of smirk that if you saw it in a bar, a fight soon broke out. His name was Lt. Kurly, but I would soon learn the men called him "common Kurly" "Another damn Yankee," he smirked. I didn't know what to say but was saved by Lt. Mosley pulling up another chair and telling me to sit down. My Lt. gave a short lecture on what he expected and instructed me to start off by being the first one working, and the last one to sit down. "Learn where everything is on the wagon first, and then learn the tools and equipment on the Engine." He finished and told me to go back to the kitchen and meet the rest of the men.

Back in the kitchen some of the original 7 guys had gone home, while more had appeared. This group was a little friendlier but they still took some verbal shots to see how I'd react. A little after 0700 the box opened up and a series of quick electronic beeps sounded. The speakers on the firehouse walls boomed as the dispatcher said, "Units 1-5-6-13-10-42-41-51 respond to 1200 East Marshall Street, M.C.V. Main hospital, fire alarm from the 12th floor duct detector." I took my place on the wagon as a captain with snow white hair emerged from the bedroom, you could see he was still half asleep, wearing his night pants and already had a cigarette in his mouth. His hand was in his shirt pocket as he searched for a lighter. He stopped walking when he saw Lt. Mosley was in the wagon and waved to him as he turned around and went back to bed. Here it was less than an hour at work, and before my official shift started at 0800 hrs. I was already going on my first run. We responded the 15 blocks to M.C.V. and upon arrival we could see 1's wagon parked at the front door. 1's Engine was a half a block back and was being hooked up to the hydrant by two men.

The Oren-Pirsch tractor trailer of Ladder 2 was parked in between them. We pulled up and our engine turned the corner into 12th St. and put in to the hydrant. We hand laid a two and a half inch line into the sprinkler connection.

Other fire trucks arrived and we all waited by our equipment for word from the firemen from 1 & 42 who had gone into the building to investigate. The dispatcher came on the radio and said, "By authority of unit 42, all units at M.C.V. can 10-22." 10-22 was our code to return to service. In 1982 only Chief Officers had portable radios so when 42 got to the 12th floor and discovered no fire, they had to pick up the phone and call the dispatcher to tell the rest of us on the street what they needed. We unhooked the line from the sprinkler connection and folded the hose properly onto the back of the engine. We arrived back at quarters just before 0800hrs. Soon the 8 o'clock gong hit and it was time for line-up. We went to the front of the apparatus floor and lined up by company in front of the trucks. 6 Truck's men on the left and 5 Engine's men on the right. Lt.s Mosley & Kurley came out carrying clip boards and read us the directives and orders that had come in over the 4 day break. They introduced me and checked our uniforms. After being told it was stove day, we were dismissed and began our daily housework.

Two companies shared the firehouse on Leigh St. But 6 Truck was only there until their new firehouse was constructed about 25 blocks to the North. They were normally quartered with 14 Engine but sometime in 1980 a large part of the ceiling collapsed in the old firehouse on Hawthorne St. revealing major structural problems. The old firehouse was condemned and had to be torn down. 14 was sent 10 blocks further north to the quarters of 16 Engine, and 16 was sent to share 15's house. I know reading this sounds confusing but it was done because 14 was a two piece engine and wouldn't fit in with 15, hence the musical firehouse rotation. Each company at 5 & 46 had one captain, two lieutenants, and 15 firemen. This totaled 36 men assigned here. There are three platoons, "A" "B" & "C", these shifts cover all 365 days per year 24 hours a day. The normal workweek averages out to 56 hours per week. Each shift has an officer and 5 firemen per company. The captain is in charge of the entire company but works only one shift at a time. After 16 weeks the officers rotate to the next shift, i.e. the A officer goes to B shift, B goes to C, and the officer from C shift goes to work with A. For example, when I was assigned to 5, A platoon had the captain, we on B platoon had Lt. Mosley, and C platoon had Lt. Pulliam. This system allowed the captain to work with each of his men, and helped deter favoritism by the officers. 6 Truck, or unit 46 their radio identification, was only supposed to be with us for a year, but stayed there about 3 years due to budget problems and construction delays.

The Richmond radio system used Company I.D. numbers that were a hold over from the pre radio and telephone days when alarms were transmitted by coded telegraph system. There once were red fire boxes on almost every street corner in the city, each box had its own I.D. number. Citizens wishing to report a fire could go to the nearest box and pull a lever that would activate a spring wound mechanism a coded wheel would open and close the circuit and transmit the box number to the fire alarm office. Let's say a box was pulled at Madison and Main Streets. The coded wheel inside the box would open the circuit 2 times then a pause then open it 9 times. A long pause before it repeated itself again to ensure the fire alarm office got the correct number. The operator would pull the box card for 29 from a drawer and could read its location and fire units assigned. Usually the closest companies would be listed for this box, in this case Engines 6, 4, Ladder 3 and the 2nd Battalion. The card also listed companies for additional alarms. In each firehouse were big brass gongs. The man at the watch desk would also pull the 29 box card and yell out if they were due to respond. In most houses this wasn't necessary as the men knew their boxes and would be headed for the equipment. Remember, in 1885 there were no phones or radios, so once out of the firehouse the horses had to go all the way to the box to find out what they had. The first unit at the box would investigate and open the box with a key. Inside the box was a telegraph key, and using a code the company officer could let the fire alarm office know what he needed, such as a second alarm, one more truck company, etc. Since it was necessary to know who was requesting more equipment each Fire Company received a code. Engines were numbered 1 through 25, so all an engine had to transmit was its own company number. Ladder companies were assigned 4 plus their number, 1 truck was 4-1, 2 truck was 4-2, and so on. Battalion Chiefs were 5 plus their number, first Battalion was 5-1. Let's say Truck Company 5 wanted a second alarm at their location. The officer would go to the box, open it with his key, send the following signal, 4-5-2-2 pause for five seconds then pull the box. The fire alarm operator would know that Ladder 5 is requesting the second alarm. And send each firehouse a 2-2-plus the box number. Once the fire department went to computer-aided dispatch in the early 1970's they stayed with the old system of numbers as it worked well with early computers

    Herman Brawn was the senior man on our shift, he was a black man in his mid 50's Over the next few years he would be instrumental in keeping me out of trouble. Herman had been a city fireman since 1960. When he was hired he was assigned to 9 Engine at 5th & Duval Sts. 9 Engine was the all black house when firehouses were segregated. He told me to grab a broom and we swept the apparatus floor together. We talked the entire time we worked. He explained to me how things worked at our house and was really telling me his expectations in a very nice way. After completing the floor we went to the kitchen where a couple of guys from 46 had already started taking the stove apart. At line-up when they said to clean the stove I figured we would wipe it down with a washcloth and clean the oven with brillo pads. Stoves in city firehouses were the heavy-duty industrial models like you see in restaurants. Each Friday they were taken down to their shell and scrubbed clean.

Dillard Tupp and Duke Smith were the other two firemen on duty that day, Tupp was a black man in his 50's and Duke also black was late 30's or early 40's and Gene-O Crump was a white guy 38 years old and off on vacation and wouldn't be back for two weeks. I had been warned that Gene-O was somewhat of a troublemaker. Lt. Kurley came into the kitchen as we finished the stove; he picked up the coffee pot and found it empty. I quickly learned that it was my job to keep the coffee pot full. While I was making a pot of coffee the Lt. started filling me in on Gene-O. "Have you met Crump yet, rookie?" he asked. "No," I replied. "Man are you in for a treat…let's see, its 0930, he's probably serviced three women and drank a fifth of Jack Daniels by now." I was really starting to wonder about this Crump guy.

At about 1000 hrs. The loudspeaker announced "CHIEF IN THE HOUSE!" Everyone on duty assembled in front of the apparatus and lined up again. Battalion Chief Frank Michaels said a few words and dismissed everyone but me. He asked me to come with him and led me into the company office. Chief Michaels was an older white man probably in his late 50's. We sat down and he gave me a five-minute lecture on how he expected me to conduct myself. Then he welcomed me to his battalion. Before we were finished talking, the alarm sounded again for smoke in the building in the 400 block of East Franklin St. Coming into the block we took the hydrant at 4th & Franklin and laid a hoseline taking a position behind 6 Engine. Lt. Mosley led us into the building taking our standpipe bag with us. We were directed to the basement and found a moderate amount of smoke there. We continued looking for the source of the smoke, with firefighters from other companies. Someone from 1 Truck soon found a pile of rags smoldering in a closet and pulled them out into the hallway with his hook. The rags were extinguished and Chief Michaels released everyone except 6 & 41. We went back to our apparatus and had just replaced the standpipe bag, and hose when the radio started squawking again. It was another full assignment in our district. Lt. Mosley called the dispatcher on the radio and placed us in service, and requested to run the alarm. Other companies did the same and soon we were leading a parade of fire trucks towards Saint James St. for the apartment fire. At St. James St. we found food burning on the stove in a three story concrete housing project. As with Franklin St. we all had our SCBAs on but no one put their facemask on as we went from room to room opening windows. 6 Truck came in and put a fan in the window. As we picked up our hose for the third time before lunch, I looked around the housing project and realized that I was the only white guy within sight. It wasn't something that upset me but it was probably the first time in my life I'd been in that situation. I knew I was going to love this job. That old saying is true, "do something you love, and you'll never work a day in your life."

Odie Denison cooked a huge lunch, after which most of the guys sat on the sofas in the kitchen and went to sleep. It was against fire department policy to occupy a bed before 1900 hrs. So the men learned to sleep while sitting in a chair. I kept busy by going through each of the tool bins on the wagon, which in fire dept. lingo are called compartments. The wagon carried all kinds of hose fittings, nozzles, air masks, rope and other essential items for doing the job. One item we carried I couldn't figure out was a large jar of Vasoline. I went to the kitchen to ask someone but they all looked too comfortable to disturb. Just as I finished checking all the tools the 4th run of my career came in, 306 N. Lombardy St. for the House fire. We were third due engine which meant we would probably be the third pumper company on the scene. 10 arrived and reported smoke from the second floor and two minutes later we pulled up to see a row of two story brick houses with smoke coming from the windows of a house in the middle of the row. Smith pulled the attack line and Tupp and I followed each pulling more hose with us. When we got to the front of the house Lt. Mosley told us to "hold up." Standing fast on the sidewalk he said "Hey rookie, come with me," and we entered the house and went up the stairs dodging the hoseline lying on the steps. At the top of the steps the smoke was thick but once again no one had put their facemask on so I wasn't going to be the only one with a mask on. The fire was in the back bedroom and # 10 already had the flames knocked down. A smoldering mattress and recliner chair had to be carried outside for final extinguishment. I grabbed one side of the chair and a fireman from 3 Truck grabbed the other side. Coming down the steps I had the high side and the smoke was making me cough and gag. The chair got stuck for a few seconds and I thought I was going to die. Being a city fireman was going to be a tough job, But I wanted to be the best fireman I could be! We got it outside and threw it on the sidewalk and the guys from 6 started pulling it apart to douse the deep-seated fire. I blew my nose and black snot came out. I could feel my lungs burning in my chest. I was thinking to myself, 4 runs and 3 exposures to smoke laden buildings, I decided I would start putting on my mask.

That evening after dinner we were standing in the open apparatus doorway talking and watching the world pass by. Herman was standing out at the curb line and I didn't notice, but every time I moved, he moved with me. As soon as I stepped out of the doorway, I got the feeling that I'd messed up. Herman was looking up at the roofline and when I looked up I could see a big slug of water in mid-air, it seemed to come at me in slow motion, but before I could move out of the way it hit me like a wave at the beach. Everyone else was in on the joke and laughed at my expense. Odie and Charlie Bolton from 6 Truck had been waiting on the roof for 30 minutes for me to step outside. I walked back to my locker with my shoes going "squish, squish" with every step I took. Being soaked with water is a right of passage for rookies in the Richmond Fire Bureau. I changed my clothes and put my wallet in my locker, I quickly learned you don't carry a wallet full of pictures in a city firehouse. I got soaked everyday for about a month and remembered in drill school Lt. Beasley telling us not to get upset when things like this happened, for firemen look for something that gets to you, once they find your weakness, you'll never have any peace.

Gene-O came in to work on about my 5th tour of duty. I'd been around firehouses long enough that I didn't need much coaching when he pulled up in a raggedy old Pontiac G.T.O. convertible. The top had more duct tape on it than top. It was a nice primer gray with body putty and rust holes everywhere. Herman suggested I go up to Gene and offer to buy his car. I knew exactly what he wanted and I introduced myself to this dude I'd heard so much about. "Hey Gene, nice car want to sell it?" I asked. "Rookie you don't make enough money to buy this car" he replied." "Sure I do, here's 10 bucks!" This comment got big laughs from everyone there but Gene, and he I got off to a rough start.

Gene was in charge of the Company that day because Lt. Mosley was on vacation. Around lunchtime we were sent to Interstate 95 Northbound at the Belvidere Toll plaza for a camper on fire. Turning out of the firehouse we could see the column of black smoke in the air and the recreational vehicle was fully involved directly under the toll plaza. Gene called for a water tanker truck because there aren't any hydrants on the interstate. We extinguished the fire with minimal extension to the toll building. Directly from there, and with an almost empty booster tank, we were sent into Gilpin Court for an apartment fire. Gilpin Court is the largest public housing project between Baltimore and Atlanta. It was an area notorious for drugs, crime, and shootings. Arriving on Hill Street we found a 2-story building with a moderate amount of smoke coming from the 2nd floor windows. Tupp took the booster line and made it up the stairs and knocked the fire without an air pack. I stopped at the compartment and quickly donned an airpack and passed Tupp on the steps, he was gagging and coughing and when I got up to the room the fire had been in, I could see he had it knocked down. I grabbed the smoldering mattress and brought it down the steps and threw it outside. I noticed getting strange looks from the older guys, wearing an s.c.b.a was uncommon for mattress and chair fires, but I decided that breathing clean air was better for my lungs, and the fresh air was free. It didn't make any sense to me to tote the heavy air bottle around on my back and not use it like so many of the older guys did. Back at the station I could over hear Gene on the phone with someone, he was bragging about what a "Firefighting motherfucker" his rookie was. It confused me as we only had a Winnebago and a room fire in the projects. I remember thinking that if this is a big deal to the "veterans" that maybe this fire department isn't what I thought it was. That thought was short lived however.

   I'd been in the company for three weeks. It seemed like everyday we had a mattress or chair fire. The small room fire on Hill St left me confused. Up until now we hadn't had the serious fire in a building that I knew was coming sooner or later. September 22 was just 4 days after my 20th birthday. It was a bright and sunny Indian summer day. Shortly after nightfall, the box came in for 421 N. Boulevard for a building fire. The assignment was 10-12-18-5-43-47-52. We were running fourth due engine company. Normally the forth due doesn't see much fire but Lt. Mosley had a trick up his sleeve that would demonstrate what a great leader he was. Coming down the Blvd. he directed the wagon driver to pull into the alley. We laid our hoseline from 10's engine and arrived behind the massive 4 story apartment house with a couple of rooms on the 3rd floor venting fire into the alley. Using the rear fire escape we broke down the back door and had the fire knocked down before the other companies could find their way back to us.

   Each day Lt. Mosley would call me out to the apparatus and point to a compartment door and ask what was inside. If I missed even the slightest tool he would frown and tell me to do better. Just the way he stayed on me made me want to impress him. Soon I knew both engines and where everything was kept. He wouldn't tell me what the jar of Vasoline was for but just smiled like everyone else I asked and told me you'll find out soon enough. When he couldn't catch me on the equipment, he started asking Street locations. I stayed in the maps at work and rode around in my car on my days off. One day I was cruising through Gilpin when a policeman pulled me over. "License and registration," he said. After checking my license over his radio he asked me what I was doing in this neighborhood. I explained to him I was a

City Fireman and was learning my streets. He looked at me in that way the police look at you when they don't believe what you're saying. "I suggest you ride this area on your fire truck, son," was all he had to say before releasing me. At work the next day I was telling the guys about getting pulled when Tupp chimed in, "We call that driving while black kid, now you know how us brothers feel."

A couple of weeks after the fire on the Blvd., we were sent to a fire at the Earle Hotel. The Earle was a 4-story flophouse located at 10 W. Main St.. It catered to the bottom rung of society, winos, druggies, prostitutes and all other people down and out. You could rent a room there for $25.00 a week, you could also catch something there bleach couldn't wash off. The box opened at 0026hrs the morning of November 8th. The assignment was for 6-5-10-12-1-41-43-53 for fire on the 4th floor. We laid-out from Adams & Franklin and could see fire showing from above the roof towards side 4 from the street.. The Earle didn't have a fire department standpipe and we helped 6 stretch an inch and a half hoseline to the top floor. Once on the top floor it was clear as a bell, I was with Mike Hale who had been in drill school with me and we looked at each other in disbelief. We went down the long narrow hallway that made a "T" shape at the rear of the building. I went to the fire escape door on side 4 and when I opened it you could see heavy fire blowing out 3 windows in the very front of the building. We hustled back up the hallway, but Pinky Parker from 10 was already through the doorway and was putting a knock on the two rooms of fire. It was lesson I would never forget; two rookies on the initial line having their fire taken from them by the veterans. Pinky was a fantastic fireman and we would race each other to see who could get water onto the fire first. Many times if my line was filled first he would drop his and help me, and vise versa.

I was living my life long dream and enjoying every minute of it, but it soon became apparent that I was very lucky or blessed too. We were first due at a daytime fire in a vacant 3 story house in the unit block of West Marshall Street. The fire was located in the "English Basement" and on the parlor floor. An English Basement has at least 50% of its area above grade level. I had the pipe and we started at the bottom like you're supposed to. 6 Engine and 1 truck were working above us. I was pushing the line in when the floor above me collapsed, dropping tons of stuff all around me. I wasn't touched by any of this but had Donny Viar from 1 truck land directly in front of me. My hoseline was pinched off by the floor and I was separated from the rest of my crew. Fire was burning all around us as I grabbed Donny and picked him up and tossed him out the window at chest high level, then climbed out behind him. The rest of 5 engine backed out of the building and we regrouped. Not a single fireman was injured. It was truly a miracle. The next work day Chief Michaels called me into the office and wanted to hear my version of what happened on Marshall St. As I told the story he was smiling from ear to ear. When I finished he told me that although I'd only been here a short time, I had already distinguished myself, he was putting me in for an award. I was humbled but I told him I wasn't hunting for medals but was lucky to be in the right place at the right time. I also felt lucky not to have been buried by the tons of debris that fell. Chief Michaels then told me something that I will never forget. "Rixner you're the kind of guy young boys want to be, and old men wished they were." For the first time in my life I was speechless.

- Chapter 7 -

The Captain

Every fireman remembers his first Captain, good or bad. I was lucky enough to have a good one. His name was Richard Emerson. He was an older white guy in his 50's. He had been in the Richmond Fire Bureau since Christmas Day 1960. He was also once a member of the elite "Flying Squad." The Flying Squad, or Unit 31 as they were also known, was created in 1965 to handle the odd jobs and the tough fires that defeated other companies. The captain rotated onto our shift in November. He was the same height as me (about 5'8") and had a thick mop of snow-white hair. He was going through a nasty divorce at home and Duke had warned me that he was miserable to be around.

Every morning when we came in for duty, I always checked over the equipment on the apparatus. Being in the house only three months I started to notice a pattern. Every time we relieved the "A" platoon, the 5 SCBAs that we carried would be filthy dirty. More often than not, half the air bottles would be low; face pieces had debris in them, etc. I would clean them, adjust the straps, change the bottles and make them ready for action. One thing nobody likes is a snitch, but by the end of November I thought maybe it was time to say something. I approached the boss and gently advised him what was going on. He took a long drag on his cigarette and bluntly put me in my place. "Look here rookie, you've only been here a couple of Months and you're complaining already. Just fix the damn problem and quit complaining or I'll find you a slow engine house to work in." I had just been taken to school. I was doing all the unpleasant jobs that rookies are expected to do, and taking

Jake Rixner

on extra projects to keep busy during the day. The captain kept a tight rein on me, but all that would change November 30, 1982

The day started out like any other day. We had what was becoming familiar, the normal automatic fire alarm runs, a mattress fire, a food on the stove job. About 2100 hours, we were sent on a house fire a couple of blocks from quarters. We were going in with 6, 41, and 51. The house at the corner of Price and Jackson had fire on the first and second floors in the rear. I had the nozzle and the captain had me by the collar. We made a push into the rear of the building and knocked the fire down and proceeded upstairs and did the same thing. I could tell by his smile and nod after the fire that he was pleased. The other thing I'd noticed was that he was a tough old guy. He never put on a SCBA and in fact was smoking a cigarette while directing us through the fire. The weather was cold and the frozen hose was rolled up and when we returned to quarters we washed it and placed it on racks to dry. We kept a full set of hose on a cart and reloaded the same number dry of sections back onto the wagon.

Duke and Tupp told me that the Captain was pleased. The coffee tasted good and just holding the warm cup in my cold hands felt great. I went into the bathroom and got some dry clothes from my locker and got a nice hot shower. Returning to the kitchen, some of the guys had gone to bed while the captain and Gene-O were smoking cigarettes and drinking coffee.

About 0300 hrs a full assignment came in for Storer Hall on the Campus of Virginia Union University. The assignment that night was 10, 5, 6,14,43,46 and 52. Storer Hall was a 4 story masonry dormitory building about 50'x200'. It had a staircase on each end and a long hallway down the center of each floor. We took a hydrant on Lombardy St. and laid about 800' of supply hose to the building. Upon arrival we could see heavy black smoke coming from the second, third, and top floors. We could also see about 100 faces hanging out at every window on the 2nd, 3rd and 4th floors. 10 Engine was already on scene and getting a hoseline ready and since the building was so large We went over to 10's wagon and helped them with their line. I ended up with the nozzle and every fireman from 10 & 5 was pulling hose so we could reach the seat of the fire. 3 Truck was throwing ladders against the building as fast as humanly possible and students were climbing down without assistance from firemen as they went back to their truck to get another ladder.

This was the fire I'd been waiting for my entire life. A serious fire was burning somewhere inside this massive brick building and possibly hundreds of young lives were at stake, and the question was, would we be able to stand up to the challenge? We went up the stairs and opened

48

the door to the second floor hallway and thick black smoke boiled out under pressure. As we donned our face pieces, I noticed my captain had a mask on. In the weeks I'd worked for him this was the first time I'd seen him with a mask on. The other man on the line with us was his older brother, Bernard Emerson, the Captain of 10. We started down the hall on our hands and knees because it was too hot to stand up. We could hear primal screams coming from the pitch black in front of us. This was the first time I'd heard anything like that and I wanted to drop the line and run for the sound. After all we were city fireman, that's what we're supposed to do, right? I asked Captain Bernard who was right behind me and he told me to stay on the line. As we went further into the darkness the screams slowly stopped, I was sure whoever it was dying. About 120' down the hall the heat was becoming greater that anything I'd ever felt before, and then to the right I could barely see the dark orange flames. 10's captain saw it at the same time and instructed me to open the nozzle and "HIT IT!" I opened the pipe and we inched forward. The sound of the water hitting the fire was like a thousand red hot woodstoves being splashed with water. The flames darkened down and we couldn't see anything when the steam hit us like a sledgehammer. I could feel my right shoulder hit the doorframe and another wave of steam burned my ears right through my protective Nomex hood. Times like these are incredibly stressful, your mind is racing and every ounce of common sense tells you to get up and run away, but you force yourself to continue forward. You're also afraid that what might be above you could fall and crush you, because you can't see anything. We proceeded forward through and over piles of burnt junk until we hit the rear wall of the room. We groped around like blind animals until we could figure out we were in a room about 16'X12'. Warning bells on SCBAs started to ring, meaning we had about 5 minutes of air left to reach fresh air. The dilemma was, where was outside? This room didn't have a window like normal dorm rooms and we had to follow our hoseline back to the hallway.

Once outside we could see all 212' of ground ladders from both 3 & 6 Trucks were placed against the building a total of 424' of ladders. The windows were now empty and I wondered if everyone made it outside? The two Emerson brothers went over to the chief who was directing the operation and reported what conditions were like inside. My ears were stinging like they were covered by hornets that continued to sting, I wanted to sit down and rest. My captain came over to the wagon and told me to change my air bottle because we were going back in. As we quickly changed bottles, I noticed steam was coming off of both of us. We went back to the fire room and the smoke had cleared, and lights had been placed to help us see that this wasn't an ordinary dorm room. It was the supply closet that was full of unused mattresses, chairs and other furniture. It also had no windows. That explains how we got burnt. When we opened the nozzle on the brick oven, the steam had nowhere to go but back on us. We overhauled the room and started to pick up our hose line. Every glass light fixture in the ceiling was melted and distorted. Some were hanging down three feet, twisted like stalactites in a cave. There was heavy smoke and heat damage to the entire floor, but the miracle was, not one fatality. What had the potential of being a massive loss of life had been turned into a massive rescue. We passed the test.

Back at quarters the sun was coming up as we brought back our second load of frozen hose of the night. Thankfully the next shift would be coming in to clean up this mess. Captain Emerson examined my ears and they were already covered with pus-filled blisters. His brother called and I could hear him talking about my ears, "Injury report?" "Just put down small burns to the ears caused by me putting my cigarette out." I had to chuckle, this was one tough old bastard.

The next shift I noticed a change in the captain's mood, no longer was I that pain in the ass rookie, I was now a full-fledged member of the team. The other guys noticed it too and from time to time someone would say, "Get the rookie to do such and such." The captain stuck up for me and said, "Do it yourself, leave the kid alone." Man it felt good. We worked every other day from 0800 hrs to 0800 hrs the next morning for three workdays and then had a four-day break.

Since I was living alone and didn't know too many people in Richmond yet, each four day break I'd drive the 100 miles to P.G. During the winter of 1982-83 it seemed like I was a magnet for working fires. Arriving at Prince George's 28, the guys would cheer, it seems that the five-day period I was in Richmond, P.G. 28 would run their normal calls, but not have any serious fires. I know it sounds sick, and no real fireman wants to see someone lose their property, but if there is a fire, a real fireman wants to be there. The best way I can explain it is if you're going to be on a sports team, you don't want to be on the bench.

Arriving at 28 in mid-December, the pattern for fires following me was known. Guys who didn't normally hang out that much at the firehouse would ask, "When is Rixner coming back?" And they would mysteriously be at the engine house when I got in. It was great because we were putting three full crews of fireman on the street for every call. The street assignment (2 engines & a truck) came in about 1800 hrs for the house fire in 30's area (I think it was in the 4800 block of 66th Ave. Engines 30-9, Truck 28.) We dumped the house (running the Engine

Company extra) the truck leading the way as it was due on the call. I had the pipe on the wagon, which meant I was in the bucket. Bolo our paid man was driving the wagon as we sped down Annapolis Rd. Sully was in the seat and we were getting dressed when I could see the truck had stopped suddenly in the middle of the road for something. Bolo was leaning over looking at Sully who was looking down at the map book and didn't see us bearing down on the tillerman's seat at the tail end of the truck. The tillerman had his head turned around and the look on his face was something I'll never forget. A collision was imminent I smashed my fist into the glass between the cab and the buckets so hard the glass shattered into a million pieces and Bolo instinctively turned the steering wheel hard to the right. We missed the truck by inches and went over the curb and came close to hitting a tree before he collected the truck and continued up Annapolis Rd. I'll never understand how we missed the truck or why we didn't blow out the tire when we hit the curb, but we didn't.

Arriving on 66th Ave., there was a three-bedroom ranch style house with one bedroom burning. Fire was blowing out one window, and 30's line was through the front door. I grabbed the 150' cross lay from 281 and headed for the front door as Sully ran to catch up. We stopped and put our face pieces on by the front door and I could see the lineman from 30 lying on the floor about 5' inside the door. Our line didn't have water in it yet and when we crawled up next to 30 Sully asked the kid, "What's wrong?" This poor kid's eyes were wide with fear and he was alone on the hose, so I reached over without saying a word and took the line from him and crawled down the hallway to the burning room. I was doing this so often for the past couple of months that it was getting too easy.

After knocking down the flames I crawled towards the window to vent. An inch and a half diameter hose shoots out 100 gallons per minute, but it also fans a lot of air. To vent with the hose line you open up the pattern into a spray stream and fill the window opening with mist. You can feel the smoke and hot gases being pulled out of the building by the air current. As we advanced towards the window I kneeled on something that I thought was a large pillow, it was lying next to the bed and I crawled through it and up onto the burnt mattress and stuck the pipe towards the window. When the smoke cleared somewhat and the truck crew brought lights in we could see that both Sully and I had crawled on and split open the badly burned body of an elderly woman. Her abdominal cavity had split open and we had blood and body fluids all over our turnout gear. We went outside and decontaminated ourselves the best we could, but what was just another easy, fun fire, had turned tragic. I have learned over the past 32 years, there are no routine fires, and even when you're going into enough jobs to become comfortable, every time you crawl inside a burning building, you're taking a big chance.

Back in Richmond for my first of three 24 hour-a-day tours, I was sitting in the kitchen talking to Captain Emerson. I'd been brooding about the elderly lady and had nobody to talk with about it. I didn't know if what I felt was normal. The logical side of my brain was telling me that she was dead from the flames long before I'd touched her, but my emotional side wondered if I had caused her death, I also wondered, if I was such a hot-shot fireman, why didn't I realize it was a body and not a pillow? I waited all day until the time seemed right to bring it up; the room had to have just him and me there as I didn't want to become the butt of a joke if other firemen heard about what I'd been thinking. I started telling him about the fire and before I could describe the woman or even mention the death, the captain said, "Hey kid, lets talk about something else besides fire department stuff". Oh well, I thought, I'll just deal with this myself.

Christmas came and went and right after New Year's on our first workday of 1983, Henry Pollard from "A" shift asked me if I would come in early for him. I got in at 0500 hrs as he asked. Henry was flying off to some exotic locale on business. Henry was a big wheel with the American Softball Association and was always jetting someplace on his four day breaks. After he left I was reading the morning paper by myself in the kitchen. The paper had a story about how there were only 5 fire deaths in Richmond in 1982. Chief Lewis cited the city's willingness to buy good equipment, and the use of smoke detectors as the reason for the low death rate. Before I could finish the article, the box opened up, "Engines 5-10, Ladder 43, Chief 52 respond to 1104 W. Marshall St for a house fire. The time is 0525 hrs." On the wagon I was working with the A team and we could see smoke coming from the third house from the corner of Hancock & Marshall Sts. We put the engine in at the hydrant and took the inch and a half directly off of it and just parked the wagon out of the way. Advancing up the steps, we found a simple mattress fire on the second floor. Just a simple $100.00 fire except Mr. Willie Coleman was on that mattress and on fire. We could see the outline of his lifeless body as we waited for water. Glenn Gran made an attempt to pull Willie off the bed but couldn't. When water came it took 15 seconds to knock the flames down. We examined Willie and Lt. Pulliam told us not to disturb anything, as the fire investigators would be coming. Taking off our masks you could smell that distinct odor of burnt flesh, just as I smelled it in Maryland a couple of weeks ago. I couldn't help but think about the article I had been reading when the call came in but it turned out to be a bad omen for the next two months to come at 5 Engine.

That same evening we were just sitting down to supper when we heard one of the assistant Fire Marshals report heavy smoke in the area of Brook & Clay Sts. "That's only one block up!" Herman said, and we all put down our utensils and headed for the apparatus floor. We could see the smoke before we left the firehouse. Coming up Brook Rd. we could see flames pouring out the front of a row of two story brick houses on Madison St. I didn't know it at the time, but it was going to be another test of courage for us. We laid out from the hydrant at Brook & Clay and when we pulled up fire was pouring out of the first floor of 405 and spreading to 407 & 403. Flames were impinging on the cars parked at the curb. A blood-curling scream was coming from behind a wall of smoke and flames. I stopped at the compartment to get a SCBA when the captain grabbed me and we headed for the front door wearing only our coats, helmets, and boots. We ducked under the flames and approximately 10 feet inside the door was Mrs. Tilley Brown. The front room was fully involved and the foyer we were in had flames rolling across the top of the ceiling and lapping out the front door. The heat was the worst I'd ever felt and breathing was impossible. The sight of Mrs. Brown was something I'll never forget. Her clothes were completely burned off and her head was bald, her hair having been completely singed off. I grabbed her under her knees as the captain went 5 feet deeper into the furnace and grabbed her under her torso. When we lifted her skin came off in sheets and I dropped my end. The captain calmly looked at me and said, "Come on kid," and I made a second attempt to pick her up. We carried her out to the street and Tupp came up and started first aid. I will never forget the look on her face as she stared at me the entire time. Engine 6 arrived and their wagon nosed in facing ours. They had deployed a line while the captain and I were making the grab and I took a position on the hose with Mike Hale and Lt. Bridges from 6. We advanced into the foyer on our knees as the hose stream darkened down the raging flames. I had my face to the floor trying to suck what little oxygen existed when I heard Mike through his mask say, "Jake, I think I found someone!" I crawled up to find Mr. Anderson Brown, curled up in a fetal position, his body was completely roasted and as I started to grab him to take him outside, Lt. Bridges ordered me to leave him where he was. We advanced past the body and at the doorway to the dining room we could see we basically had a one-room fire. I went out to the street and snuck behind Truck 1 and vomited into the sewer grate. I remember hoping nobody saw me vomit because I didn't want to appear weak. The investigators came and determined that the 79-year-old Mr. Brown had tried to move a kerosene heater while still lit and accidentally turned it over starting the fire. The rescue with Captain Emerson was the craziest thing I'd ever done at that point in my life and thinking back at the smoke and flames I realized that I was working with a special breed of men.

In the next couple of days we would check on Mrs. Brown at the burn Unit at Medical College of Virginia (M.C.V.). Mrs. Brown was burned over 90% of her body and the prognosis wasn't good, but she hung in there day after day. I think it was during this time that I really grew up. Up until now I had always viewed firefighting as something fun. It was the ultimate adrenalin rush. At 20 years old I now doubted whether I could work 40 more years in the Bureau until the age of 60, which was our retirement age at that time.

We had working fires almost every day and a couple of times a month, we would come to work and the apparatus floor would be empty, as the other shift was still on the scene of a fire. We took turns and drove our personal cars to the fireground to relieve the other shift. I never liked that arrangement as I've always been picky about my automobiles and the guys returning to the firehouse would always be filthy dirty and soaking wet. In short, they would leave your car a mess. I was studying the streets and hydrant locations as a new man was expected to know every street in his first alarm district. I remember running the daily M.C.V. alarm and the captain asking me if I felt comfortable driving back. "Sure," I said, and was proud of the fact he took notice of my hard work in learning our area.

Coming down Brook Rd. about two blocks from quarters the beeps came in over the radio. "5-6-41-51, 626 Judah St. for the house fire." "Do you know where that is kid?" the Captain asked. "Yes sir, down Leigh to the first left, hydrant on the corner of Duval & Judah." "O.K. take it," he said as he turned on the red lights and began to sound the siren. Every fireman soon learns that the siren is just a loud deafening noise but this time it sounded like the Calvary to me. "We are coming to save you," it was saying. I turned into Judah and the last house on the right had smoke coming from it. We laid out from the hydrant and made quick work of the room and contents fire. I'll never forget how proud I was during that time period. It seamed like the tests were coming at me everyday, and I kept meeting the expectations of my captain and my co-workers.

On January 14, 1983, at the 0800 hours line-up, Captain Emerson was reading the roll call. "Drivers for today are: Wagon-Rixner, Engine-Crump, Truck-Baker, Tiller man-McEntire." So there it was I was officially a driver at Engine 5. Man what a feeling, I'd been keeping up with other recruit school classmates and no one else was turned over to drive yet.

My first day driving was just another normal day; we went to M.C.V., a mattress, and a false alarm in Gilpin Court. Most of the guys were sleeping in the kitchen chairs and I had the wagon partially out the front door putting fuel in it when the box came in for 1834 W. Grace Street for the apartment fire. The assignment was 10-12-5-18-43-47-52. I was driving down Leigh St. as fast as the wagon would run when the captain told me, "slow down kid." 10 went on scene and reported "WE DO HAVE FIRE." It was my captain's brother Bernard. It could be one room or the whole block on fire; that's all Captain Bernard Emerson ever said on the radio, in all the years I worked for either brother I never saw them get excited. I was trying to emulate them but there was one guy on my shift who would yell and scream when a run came in. It was difficult to block him out and I wanted to be the best fireman I could be.

Grace Street was one block directly behind # 10's quarters. It was probably the most unique street in Richmond. Each block was different than the last one. One block would be gentrified and all fixed up with doctors and lawyers living there, and the next a ghetto complete with winos, and derelicts, whorehouses, the works, you can probably guess which blocks we did most of our actual work in. We laid out from 10's engine and I parked the wagon directly behind 3 Truck.

In all my excitement to get the company to the fire I didn't even think about blocking the access to the ground ladders stored on the back of the hook & ladder. There was fire in the basement and when we got down there we were the third hose line in the basement. The fire was knocked down except for a natural gas main that was broken and on fire. There was a guy from 12 named Wayne Dempsey who kept putting the gas fire out. Now this is exactly the wrong thing to do unless you can shut off the gas. Natural gas is lighter than air, which means it will rise. If given a chance to collect in a pocket it will explode. The captain told him several times to stop putting the fire out but he would just yell like a wild man and do it again. Thankfully there were enough glowing embers on the floor joists and in the room to reignite the gas each time.

The captain told us to back out of the basement and we did taking our line to the first floor of the building. We found fire burning through the baseboards and walls and the guys from truck 3 were opening walls and ceilings for us. I drifted off from my company and soon found myself on the second floor. I was crawling around when somehow I got turned around and lost my sense of direction. About the time I realized I didn't know where the stairs were. Fire flashed on the baseboard. This building was full of hidden fire and here I was off on my own freelancing. To make things even more interesting, the low air bell on my SCBA began to ring indicating that I had 5 minutes of air left. I needed to get out, but I didn't know where out was! I could feel panic beginning to build up inside me when I just stopped and took a deep breath. If I was going to die in this building, I was going to die trying to get out in a methodical fashion. Just then this great big black man with 3 Truck on his helmet grabbed me and asked if I was all right. It was Bobby Nelson "Hell no, I'm not alright Bobby, I'm lost!" Bobby chuckled and said, "It's all right rookie, I got you," and picked me up and placed me across his hip with one arm and carried me down the steps.

Back on the street I was relieved and wondered if my company had missed me. At the back of the wagon I saw the captain and the rest of the guys from 5 engine, "WHERE IN THE HELL HAVE YOU BEEN?" I could tell he was upset with me. I'd been doing so well the past five months that I got a little too sure of myself. I apologized and explained that I went up to checkout the second floor. "FROM NOW ON YOU STAY WITH THE COMPANY OR I'LL PERSONALLY KICK YOUR ASS, YOU UNDERSTAND!?"

I was feeling like a heel and knew I'd screwed up when Sam Samuels, the driver of 3 Truck came up to me in front of all the other guys and pulled a ladder from the ladder truck. It only came about halfway out and lined up perfectly with the windshield of our wagon. "Hey rookie, the next time you pull up tight on my ass, I'm going to put this ladder right through your fucking windshield!" Talk about having your ego deflated. The rest of the tour was uneventful and I felt like the biggest screw-up in the world. I wanted to be the best fireman in the world, but I could see that I was going to take my lumps along the way. I just made up my mind to not make the same mistakes twice.

It snowed on our day off and the next workday and when I got into work; both pumpers had chains on them. The day started out with the usual 8 o'clock M.C.V. alarm and at 0910hrs we were sent to 1604 Hickory St. with 6-46-51 for a house fire. We pulled up in the 20-degree temperatures to see white smoke banked down so low; we couldn't even see the one story house on fire.

I had the nozzle and we went through the place using our sense of sound and touch. After the fire we reloaded the wet frozen hose as we had long ago depleted our spare hose. The air packs were frozen except two of them as all three shifts were having at least one fire per day and it wasn't unusual to have two or three.

I have wanted to be a fireman since as long as I could remember, but the fires, deaths, and cold frozen fingers and toes were like a reality check for me. I could see the other guys on my shift were suffering too. Herman and Tupp were in their 50's and you could hear them moan as they rolled out of bed for the runs.

Duke and Gene-O were in there 30's and they were suffering as well. None of them could understand why anyone would volunteer to do this kind of work and Tupp reminded me frequently that anyone who did would have to be retarded.

That night just after dinner a man came running up to our front door in the snow without any shoes on to report, "MY HOUSE IS ON FIRE!" We looked up the street where he was pointing but couldn't see anything. "Where is your house?" Duke asked. But he was so excited all he could say was "BY THE CHURCH!" Now this might be a good description in many neighborhoods, but Jackson Ward had a church on almost every block. We got on the apparatus and headed east on Leigh St. Stopping at each intersection to look up and down the street. At St. James and Leigh we could see a two story brick hose at the corner of Duval and St. James fully engulfed in flames. The captain picked up the microphone and asked for the assignment. It would bring 6-41&51 out to help us, but we would be here by ourselves for the next three minutes. I took the nozzle and we had to knock the fire back from the street just to get onto the front porch.

Once inside the front door, I swept the straight stream back and forth across the ceiling deflecting the water as we inched our way towards the back of the house. The heavy plaster ceilings began to fall and the floor shook when they hit. We "turtled up" meaning we tucked our necks low in case the plaster hit us, but it would be of little help as there isn't a helmet in the world that would protect your neck from the weight of plaster hitting it. We were in the kitchen when it felt like somebody hit my head with a sledgehammer. The plaster came down and struck me so hard I was like a wounded animal crawling around in the debris. The Captain grabbed me and dragged me outside, my neck felt like it had been run over by a truck. I remember watching 6 take their line upstairs and being pissed that we didn't get the entire fire knocked down before they got here.

After a moment or two I regained my senses and we went back inside and watched 1 Truck open the walls and ceilings. Under the heavy plaster was little wooden strips called lathe. It takes strong arms to work the long pike pole to pull down the lathe strips. After the "Truckies" had exposed the entire room we would wet down the hidden fire or hot spots they had exposed. After loading the wet, frozen hose back on the wagon we returned to quarters. It was about 9 p.m. when I opened my locker and discovered I didn't have any dry underwear or uniforms. The dirty soaked clothes I had on would have to do for the remaining 11 hours left in our tour of duty. I got into the shower and let the scalding water wash over my sore neck. I blew my nose several times before the snot didn't come out black. A good hot shower was the only respite from the long cold winter, and became a special treat after each fire.

After the shower I picked my cleanest pair of wet dirty underwear, put it back on and went into the bedroom and lay down. I always kept several extra sets of clothing in my locker, but because we had been so busy I didn't have enough. I made a note to stock more extra clothing in my locker. Lying on the bed I realized my muscles ached to the bone, I had drying scabs on both ears, snap neck, and I was lying there praying to God that we stay in the rest of the night. Before I finished my prayers the box opened up and the dispatcher sent 10-5-43-52 to the 1800 block of W. Leigh St for the house fire. Coming off the front ramp we could see a loom up of black smoke in the western sky. I thought about quitting right then and there. Once we got to the scene it was too far to walk back to quarters. I think that's the only reason I didn't quit right on the spot. Here I was living my life long dream and I honestly didn't know if I was tough enough. I ended up behind Pinky Parker from 10 and we pushed the line through the building with Captain Bernard Emerson right beside us "Good job, that's it, keep it moving," he coached us through the fire. I remember in the last room there was an overstuffed chair burning and we left it burn so that we could use it to warm our hands during the overhaul process. It was our version of a warming barrel you see on construction sites. After the overhaul was done we put the chair out and threw it outside with the rest of the debris.

After the fire and back at the firehouse kitchen, I made up my mind. I was going to stay at it, and work this job, so what if the fires didn't come according to my schedule. I know firemen in other places that would sell their first-born child to see this much work.

The captain rotated to the "C" shift in February. We were sad to see him go, but the fire duty we performed in the 16 weeks he was with us was something I'll never forget I think we ended up with 47 working fires and 9 fatalities, & a couple of successful rescues, including the 100 students 3 & 6 truck brought down over ladders. On January 26 Mrs. Brown died at M.C.V.'s burn unit. She had put up one hell of a fight, but in the end the fire got her too.

- Chapter 8 -

Fire Fighting Super Stars

I was working with some real superstars and it seemed someone was saving a life a couple of times a year. Making a grab as we call it took a lot of luck. Fire buildings are pitch black, and it comes down to being in the right place at the right time. My next successful grab in Richmond came in March of 1983. The fire came in on a drizzly, dreary day. The box opened for 100 block of West Clay St. for a house fire. The assignment was 5-6-41-51. We arrived to find a 2 story brick dwelling with heavy smoke pouring from the first floor. The dispatcher didn't say anything about people trapped and I don't know why I did it, but I entered the house without a line and proceeded down the hallway towards the rear of the building. The fire was free burning in the last two rooms to the left and I entered the kitchen to find a young black woman with her two children clawing at the back door. The door had wooden planks nailed horizontally across it to prevent burglars from getting in. The problem was it was now preventing the young family from escaping the rapidly deteriorating conditions. In a couple of seconds we would all be in trouble because the fire was really starting to take off. One child was in diapers and I grabbed her and put her in the fold of my left arm. The second child was a little older; maybe 4 or 5 and I picked him up with my right arm. The woman was sobbing and on the verge of hysteria and I grabbed her arm with a hand and started to lead her back towards the front of the building. When we got to the doorway from the kitchen to the hall the fire was rolling across the ceiling and escape would soon be cut off. She pulled free from my grasp and retreated to the back door

Jake Rixner

again and I was coaxing her to come with me. "Come on, I made it in here, you can make it out." She reluctantly followed me back to the hallway again and this time we made a dash for the street.

Out on the sidewalk 6's wagon was just pulling into the block. I set the little boy in the front seat of our wagon and when I went to set the toddler down on the sidewalk, I let go of her for just a second and she wobbled and fell butt first into a puddle making a big splash with her diaper. The lady was sobbing uncontrollably and was trying to kiss my face through my mask. I rejoined my company and followed Duke and Tupp thru the two rooms as they knocked the fire down rapidly. After the fire was over, a guy from 6 we called Radar came up to me and said, "You know that was a nice rescue you made, but next time try not to throw the baby in a puddle." Fireman will bust your chops even when they are complimenting you.

Lt. David Pulliam came onto our shift about this time and I wondered about him. It seemed like he was always playing around and didn't take anything seriously. He was a white guy in his 30's and gave me the impression that something was wrong with him. The run came in a couple of days after the Clay St. rescue and it was for an ammonia leak at the E.M.Todd Ham Processing Plant at Leigh & Hermitage Rd. It was about 10:00am on a weekday and the area was full of traffic. When we got on the scene 10 & 43 were already there. Lt. Pulliam got out of the front seat and went to the compartment, he took out that large jar of Vasoline I'd been wondering about and dropped his pants and underwear and began to cover his genitals with Vasoline in broad daylight. Standing there with his pants around his ankles he was yelling, "Rookie, come here." I couldn't believe my eyes. Now I knew this sum-bitch was nuts!!! I reluctantly went over and he explained that the Vasoline was to protect you mucus membranes from the effects of the ammonia and I was to put some on my crotch, ass, and underarms. The face piece would protect your eyes, nose, & mouth. I let out a sigh and dropped my pants, and with by-standers watching us in disbelief; I coated my private parts in Vasoline. Remember this was before the high-tech hazardous materials teams most fire departments enjoy the services of today (modesty was not even thought of at that time). We went in and found the leaking pipe and traced it back to the nearest valve and shut off the dangerous leak. Ammonia under the right conditions is explosive. The Todd plant would have one of these leaks just about annually and the worst part about it was trying to get all that Vaseline from the crack of your ass. I never imagined that being a city fireman would involve such a variety of jobs.

As I got to know Lt. Pulliam, I realized that he was one of the best officers in the bureau. He had the biggest set of balls literally (I saw them) and figuratively. He was always jumping off the wagon and running into the fire building with only a helmet and coat on. He never stopped for boots or a mask and would run in and find the fire and meet you inside and guide you to the seat of the fire.

The fire came in on a Sunday morning at 0905hrs. Our cook Gene-O was at the grocery store buying the items for the meal, so we were a man short when the box opened for 5-10-43-52 to 1112 Catherine St. for the apartment fire. I was driving the wagon and Wilbur Johns and Herman Brawn were on the backstep with Grayson Finner driving the Engine. The 10-11 &12 hundred blocks were a bit of a race between 10, and us and if you were slow 10 would beat you in there. Catherine Street came in a couple times a week and usually was a decent fire. It was another unique street because anytime you went there, day or night, it seemed like everyone was drunk. We turned from Hancock St. into Catherine when the smell of burning wood and plaster hit us. It was the smell of another job for 5 Engine. We laid-out from the corner and coming up the block laying hose off the back of the wagon we could see a 2 story wood frame quadraplex (4 unit apartment building) with the front room on the bottom left side heavily involved in fire. The fire was out the two front windows and lapping along the underside of the porch. Just before we pulled to a stop a black man with half of a six-inch high afro style hairdo came out the front door with smoke and flame right above him. He only had half the afro because the other half had been singed off. He was slurring his speech and yelling, "Willie is in the back, man. You haves to go gets Willie."

Pulliam ducked down under the flames coming from the top of the doorway and disappeared. Wilbur and Herman went for the masks in the side compartment while I wondered what to do. We normally didn't use the pump on the wagon as we would break the two and a half and put a pipe on it or break it and dump the 150' dead load of inch and a half with a wye. I had only my coat and helmet on and grabbed the 50' trash line from the side of the wagon and flaked it out and with a flick of the wrist I'd already put the pump in gear and charged the trash line.

Knocking the fire back I advanced the line into the doorway. The fire was in the front room just to the left of the front door. The afro man grabbed the line behind me and said, "I'm with you brother, I'm a Marine Corps". The smoke came down low and my marine buddy bailed out on me as I entered the room that was fully engulfed. I remember thinking "Semper Fi, motherfucker." I was pushing in when something fell on me and instantly burned my forehead. This caused me to flinch and when I did my leg hit a red hot metal chair leg burning me there too. I was confident I'd put a knock on the fire and getting burned twice and being unable to breathe, I turned around to back out as 10 & 43, and my crew was pushing their way in. Unable to escape that way, I handed the line to Herman and went forward and out the freshly burned-out floor to ceiling window opening and onto the front porch.

Out front I looked for Lt. Pulliam and found him with Willie on the back step of the wagon. They both were holding oxygen masks to their faces and it was a relief to see they made it. After overhauling I asked the Lt. if he was mad at me for using the pump on the wagon. "I was afraid with the fire coming out the front like that you would get cut-off, and we know how the backs of these joints are usually fortified with bars on the windows and such." I had only been working with him for a couple of days and didn't know if I should have followed him into the building, kept to the normal procedure of hooking up the line we laid from the engine, or using the trash line like I did. He looked at my burned forehead and leg and told me that what I'd done was fine. As it turned out the back of the building was fortified and escape was impossible. It's one of the sad things about poor neighborhoods that people have to put metal bars on their windows and doors to keep the dopers from stealing everything they own.

My forehead and leg blistered up and with my scarred ears and neck, it seemed like I worked that entire winter with bandages on some part of my body. We called them blowout patches. Lt. Pulliam was a close friend with Captain Butler in 1 Truck. Everyday we would get on the equipment and ride down to 6th and Marshall Sts. and spend time with 41. Captain Butler saw my latest battle scars and shook his head at me and went into his office, he returned a few minutes later and had a requisition form the bureau used to request supplies and other needed items, he handed it to Lt. Pulliam who read it a burst out laughing. Pulliam passed it to Eddie P. and he laughed and said, "Sure, you're right." Eddie passed it around until everyone but me had read it. When they gave it to me Captain Butler had filled out a request for Nomex skin for me. I had to laugh too.

As the weather improved, men began to take vacation days; each shift had an officer and five men. In the wintertime when no one was off, we worked with six on the fire trucks. It worked out very well because the months of December, January, and February were crazy with all the fires. The minimum manpower in each unit was four and one, meaning four firemen, and one officer. As guys began to take days off, the Battalion Chief would balance his manpower by transferring men from one station to another for the day, we called it filling-in.

The first day I had to fill-in I was sent to 1 Truck at 411 N. 6th St. 1 Truck was in the old armory building with 3 Engine. 3 had been deactivated in June of 1981 due to budget cuts, so 41 had the entire house to themselves. The supply center was in the basement of the Civil War era structure and we had a delivery van that shuttled supplies and carried about 50 spare SCBA cylinders. This truck was known as the "Service Truck" and had a radio i.d.# 91. I showed up at my normal time of 0700, an hour before we officially assumed duty. Relief was a man for man deal and as soon as your man was in you could go home. Captain Butler was working as acting Battalion Chief 3 (or Unit 53) and was already at the firehouse having reported to 17 to relieve the chief who worked last night. He would leave 17 and run the Battalion out of 41's house as it was his normal home. Fred Biltner was waiting for me at the kitchen door because I was his relief and he started badgering me before I entered the kitchen door. Fred was a weight lifter and her could make his pectoral muscles jump so that it looked like he was waving at you with his man boobs. The nurses at M.C.V. loved it and would hunt Fred down each time we went into M.C.V. for the daily false fire alarm. Fred was always giving somebody a hard time, but it was always good-natured. He told me, "Rookie you've got the service truck today." "O.K. Fred," I replied, "What do I need to do?" "First off you know you've got the air-bottles, right?" he asked, "That means you respond to any working fire north of the river, you understand?" "Yes sir," I said. "O.K. kid, knock off the sir shit or I'll squeeze your head like a pimple." Fred said mockingly looking hurt. Then he led me to the watch desk on the left hand side of the apparatus floor where a clipboard was hanging up. On the clipboard were all the requests for supplies and broken equipment that would be taken to the appropriate repair shop. The requests came in to 91 all day long but you only did what was reported before 9am and primarily took care of yesterday's requests. "O.K. kid here's how I work the truck, any fire you go to, park the truck as close to the fire as you can get, tell the chief where it's located and go find a hoseline to steal," Fred told me, "some of the best fires I've fought were on the service truck," Fred told me before he left.

I had been out at the bars the night before and still had a big head when I went into the bathroom to use the urinal. Standing at the urinal my face was almost touching the wall in front of me, "Man I must still be drunk," I thought. Later during clean up I had the mop bucket and wheeled it to the doorway of the bathroom, I stopped it at the door and when I let it go it rolled towards the opposite corner of the room. The floors were slanted so badly that you had to leave the mop bucket in the hallway. I laughed and felt much better as I wasn't still drunk, this old firehouse was just in bad shape. Eddie P. showed me the walls on the apparatus floor, there where white stains on the tan paint. "You see that kid? That's pigeon shit, it runs down the walls every time it rains," he explained. Upstairs was an old basketball court complete with bleachers. Many of the windows had been broken over the years and the pigeons had moved in and taken over. I can't help but think back of what a hazard just breathing the air in that old barn was.

About 0930 hours. I set out to make my rounds and each station I visited seemed to have a section of hose that needed repair, a broken Circle-D lamp, fan, or some other kind of tool. I was running all over town when I realized that I'd left 41 that morning with 8 stops, and now on my 15th stop I still had about 10 more to do. The dispatcher called me on the radio and told me to phone my quarters. At the next stop I called 41 and Eddie P. was some kind of pissed off. "Where in the hell are you?" he asked, I told him about the additional work and he advised me not to pick-up anything not on my list and to "Get my young white ass back here now." I didn't know it but Eddie was waiting on me to get back to quarters so he could run out for a couple of minutes to take care of some personal business. I returned to 41 and he told me to ride in his place on the hook & ladder until he got back. I put my gear on the truck, but I was puzzled, weren't we supposed to try to get all the broken equipment to the proper shops to be fixed? It was just another lesson for a new man. Eddie was the senior man at 41. I found out that he ran a social club and repaired cars on his days off. He would always be dead tired and when he spoke he mumbled so bad that only those who had worked with him for several years could understand him. He ended every sentence with the word shit. It sounded something like "mumumumumSHIT!" You could tell when he was upset because he spoke faster and louder, but it still sounded the same.

Eddie got back later in the afternoon and I still had 2 more stops on the days list so I went back out in the service truck. I was coming down North Avenue when the radio began to beep and the dispatcher put out a building fire in the Scott's Addition neighborhood. I was just about back at 41 when 18 went on scene and asked for a second alarm. I flipped on the red lights and headed for the fire. Coming up Leigh Street I could see #5 responding on the Second alarm about 10 blocks ahead of me. By the time we got to Scott's Addition I had caught up to them. I parked the truck on the corner and advised the chief where it was, got my gear on and hooked up with #5. A mattress factory was burning and the concrete building took up the entire block. We laid a line down the side street and forced a door that led into the warehouse. Gene-O was in charge and we took a two and a half inch line into the thick black smoke. The fire was burning just inside the loading dock and all of the first alarm companies were working on that. The chief had sent us to the opposite side of the fire to try to work our way in behind it. Grayson had the pipe and I held on to his air bottle as we walked through the thick smoke boxer style. The front foot probed for the floor with each step. Our line was still dry since we didn't know exactly how much hose it would take to reach the fire. It was an uneasy feeling being hundreds of feet into this black abyss without any water and I was worried the fire just might find us first.

Finally we could see a dim light ahead of us and we could also hear the other firemen working. We had reached the seat of the fire, but it was under control. I walked through the overhead doorway without realizing it was a truck loading dock, I had let my guard down and before I knew it I was falling. The fall only lasted a second and I twisted in mid-air. I landed on my hip with a big splash in about two feet of water that had accumulated. I went under for only a second and when I came up the news cameras of Channel 12 had caught the whole thing. A couple of guys jumped down to see if I was alright. I felt like an idiot again. I got up and limped over to our engine, which was put into the hydrant. Herman was at the pump panel. He asked if I was all right. "Yea, just another dumb rookie mistake," I said. "Jesus rookie you gotta slow down a little. If I'd have done that I'd have broken my hip." Herman said. It seemed like every time I was doing good, something would happen that taught me a lesson, I was still determined to be the best fireman in the world, if I didn't kill myself first., and I was starting to wonder which one would be first.

The next morning Fred came in to relieve me. He already knew about my acrobatic stunt. He was teasing me and he said "Maybe you should just stay with the service truck and pass out air bottles".

Back at # 5 the next day two newspaper reporters were there to do a story on the busiest Engine Company in the city. For years 11 Engine on Church Hill had been the busiest, but when they closed 3 & 31 for budget reasons, this propelled 5 to the top of the list. We answered 1325 alarms of Fire in 1982, and a large percentage of those were working fires. Number 5 also had an advantage of being centrally located in the North Side of Richmond. If we weren't due on the get-go, we were definitely due to respond on a second alarm. They were only with us for about four days but they saw plenty of action. The "A" shift had a fire on Catherine Street. Heavy fire was blowing out of the back of a two story wood frame house. A man was on the front porch when they pulled up, the wife had climbed out onto the porch roof directly above the man, and both were yelling for the firemen to get the other one. It was kind of comical as they both made it out and were literally standing one above the other but didn't know it. We took in a couple of fires; one at 1st & Leigh Sts. Doug Wilder owned the house. Wilder would later go on to be the first elected black governor in any state since reconstruction. I was driving the engine that day and was put in at the hydrant directly in front of the building. They stood close by and peppered me with questions. Another room and contents job in the 100 block of E. Charity Street. The story came out in the paper a few days later and they did a good job of describing life at our engine house.

Engine Co. 5 was building a reputation of getting the job done. Most firehouses will have a couple of great fireman assigned there and a couple of posers. A poser is a guy who wears the uniform, but disappears when the scene gets tough. Self preservation is usually their motive. Number 5 had a majority of great fireman. All three shifts seemed to get along great, any problems that arose would quickly be handled by the senior guys, like Herman on our shift, Everett Grimes on the C shift, and Henry Pollard on "A". That's unusual in most firehouses. During this time we had several guys from other companies who wanted to transfer to 5, this kept everyone on their toes because you didn't want to screw-up and get transferred. Not only was I living out my life long dream, I was doing it at the best and busiest firehouse in the city. Everyone should be so lucky.

The box opened up at 0501 hours. Everybody was in bed when the lights came on and the gong struck for a fire in the 3600 block of East Broad Street. 5 Engine was 1st due on a second alarm for a house fire. It was very rare to go on a second alarm in Church Hill since the Church Hill neighborhood had some excellent firemen who could handle anything a fire would throw at them; this was something new to me. This meant there would be some quality work for us when we got there. I was riding the back step on the wagon that night and as we were coming up East Broad Street, the sky was just beginning to show some sign of daylight. As we approached the fire scene, the first battalion chief assigned us to lay a line to the rear of the building. We stopped across from 36th street and took the hydrant that was right at the edge of the alley and laid to the rear of the building.

What nobody had told us as we were coming in was that numerous people were trapped in a 2 story wood frame house that had heavy smoke and flames pouring from everywhere. Another thing we didn't know was that we were assigned to take a 2 ½" line in from the rear and that involved navigating a flight of exterior wooden stairs that looked like they were about 80 years old. If we had actually been able to see the condition of the steps, we probably wouldn't have wanted to climb these steps on a good day without all our heavy gear on, let alone having 3 or 4 fireman with about 80 lbs. of gear on, tromping up these steps. When we got to the rear of the building, we dropped a couple of sections on the ground uncoupled it and added a 2 ½" fog nozzle. I positioned myself at the pipe and started up the rear stairs. When we got near the top of the stairs, we found that the smoke had banked down to the street level and we had to don our face pieces before entering the building. In my haste, and we were in a big hurry to get to the back of this building, I didn't have my low pressure hose screwed all the way into my regulator. I made it about five feet into the hallway with Tupp behind me when my low pressure hose came undone from the regulator. I began sucking hot, acrid smoke into my lungs. I tried desperately with my gloved hands to get the hose hooked back up to the regulator, while hanging on to the nozzle which was flowing 250 gallons of water per minute. It was too much for me. I couldn't breathe; I couldn't hang onto the line, so Tupp ended up moving up and relieving me from the nozzle while I backed up to the back porch behind Grayson and the rest of the guys to fix my mask. By the time I fixed my mask and got back up nobody wanted to yield the nozzle. Tupp had a hold of the pipe and he wasn't giving it up. I was kind of disappointed, but I took my place in the back of the line and with a 2 ½" line, once you get inside the building and on the 2nd floor, it is up to the backup guys to determine how far the line goes; it's a teamwork type of deal. I probably gave up an easier position for a harder-working position because now it was my job as the last man on the line to pull the hose up the flight of stairs (a charged 2 ½" line).

It was very labor intensive to drag that hose up the stairs from the rear yard and make sure there was enough slack in the line to keep it moving forward. After a minute or two, I could hear Tupp calling that he had a victim and expected him or someone to grab the victim and drag it out past me. I was surprised when nobody came by. Within five minutes, we had the entire 2nd floor knocked down and as the smoke began to clear, it became apparent we not only had one victim, but indeed we had 4 fire fatalities. The sad part was that three of the four probably could have survived with proper fire safety education or training. It was cold that night and it appeared that three people in the front bedroom, a very heavyset woman and two children about 8 and 10 years old were all fully dressed with winter coats and shoes. It seemed they had woken up during this fire and gotten dressed and they were right inside the front window of the front bedroom on the second floor, and they had dropped there from smoke inhalation and died from the heat. They were piled on top of each other. The fourth victim, the one that Tupp found, was right at the top of the steps. The first floor was fully involved when Engines 1 and 11 got there and this individual was burned so badly that we couldn't tell if it was a male or female. It was literally a hunk of burnt meat. I had never seen a person burned so badly that his/her sex could not be identified; it was that graphic.

After the smoke continued to clear, we began opening the walls and the roof, trying to preserve the dignity of the victims, by not pulling debris on them, but we did have to make sure the fire was out. We tried to save as much of the building as possible for the investigators so they could piece together what had happened here that had ended up costing four human beings their lives.

When the story came out it was a rooming house, there was a birthday party the night before and a since it was a cold night, a gentleman gone back to his room on the first floor from the party, and turned on a hot plate with some water on it to try to warm up his room a little bit. The hot plate overheated and was the cause of the fire. The most heartbreaking part was looking at the two children and the woman in the front bedroom who could have simply just stepped out onto their front porch roof and crossed over to the porch next door and since all the houses in the Church Hill neighborhood are of similar construction, they could have walked down porch roofs for almost a block if needed to escape from the fire; but the smoke had gotten to them first.

I work with superstars and have seen guys make a difference and save lives but it can be frustrating, because even with a full cast of superstars, that night, we were too little too late and there was nothing we could do. This is one part of the job that is always frustrating.

- Chapter 9 -

Big Fires

In a 2 ½ year period from November 1984 until March 1987, our shift experienced a number of major fires. By major fires I am talking about fires ranging from a 3rd alarm to an 8-alarm fire. These fires were instrumental in shaping my outlook and sparking an interest in strategy, and tactics and what is really savable, and not savable as far as fire ground command structures. I learned a lot at these fires from watching the chiefs who had the fires and how they organized their resources to mitigate the damage or to stop the ongoing damage. At some of these fires, there is nothing that can be done except to back off to a fire break such as a fire wall or vacant lot. Some fires are so far advanced upon arrival that there is not much that can be done and to let the fire take possession of what it has until it reaches a physical boundary where the fire can then be stopped. One of the things that I quickly realized was that some big fires are the result of poor building construction, over crowding or poor maintenance.

A very large, out of control fire is an awesome force of nature that demands respect. Just about every fireman at some time during his career has heard other companies responding to a large incident and firefighters as a rule will quickly gather around the scanner or radio hoping to be called next so they can see exactly what is going on. There is some voyeurism involved, plus everyone wants to be a part of the action. 5 Engine was not 1st due at any of the major fires we went to during this period of time; they were all multiple alarm responses.

Jake Rixner

One of the first big fires I worked in Richmond occurred on November 12, 1984, at The Standard Paper Company at 1st and Hull Streets in South Richmond. I remember 1 Truck had just gotten their Firebird (a 100' elevating platform/snorkel type apparatus) and this was one of the first major fires that I was able to see the unique capabilities of an articulating type boom in action. This call came in as a daytime fire and 13 which was usually first due was in the shop when the alarm was sounded. Engine Company 33 ended up being the first in on the fire which was pretty well advanced in this late 19th century brick and joist building that was a paper manufacturing plant that had been vacant for several years. Upon their arrival 33 was met with heavy smoke conditions and fire in the basement of this 150 year old building. They gave it their best shot using a 2 ½" attack line but after a few minutes, it was apparent that the fire was continuing to grow. In fact when the chief arrived, he pulled everybody out and we went to a defensive exterior attack. A defensive attack is when you look at the big picture; sometimes you just have to surround the fire and try to cut it off from spreading to other buildings. The idea of defense is that you are giving up trying to save that building and attacking the fire and instead focus on an effort to stop the fire from spreading to other structures or jumping streets. We responded on the 2nd alarm and Captain Marsh was in charge of the company. When we responded, the 4 engine companies on the 1st alarm had taken the closest hydrants, so Captain Marsh instructed me (I was driving the wagon that day) to scout out the area for the next closest hydrant so we wouldn't have to lay blocks of hose. We finally secured an unused hydrant at about 4th and Bainbridge Street and laid out all the way down Bainbridge Street. The building had some access problems including the fact that there was a canal adjacent to its south side. On the north side of the building was the James River and then flanked on the east side was the foot of the 14th Street Bridge, and on the west by the Manchester Paper and Board Company which was an exposure less than 15' from the fire building. We ran out of hose exactly in front of the fire building on the opposite side of the canal There was a small bridge that was used to cross the canal, but Captain Marsh decided that since the chief had gone to a defensive attack at that point, we were instructed by the captain to set up our stang nozzle (a large water cannon capable of flowing 1000 gallons per minute) right there on the ground on the opposite side of the canal. It didn't look like we had that bad of a shot and our engine was put in at the hydrant just 3 blocks behind us, but as soon as we began to flow water it became apparent that friction loss combined with the length of hoseline (approximately 1000') resulted in not enough water pressure and the water lobbed over the canal and hit the fire building about 5' below the window on the cinderblock wall that was showing heavy black smoke. It was kind of embarrassing because the wagon we had at the time had only a small 250 gallon booster pump and that was of no use to us. The captain was running around to find an incoming engine company that could put into the middle of our line and pump to increase the nozzle pressure which would give us a decent nozzle stream. To add to this circus, there was a small railroad siding of 3 tracks that was used to serve customers, it wasn't a mainline track so we didn't have to worry about a train barreling toward us at 60 miles per hour. While we were trying to get a decent fire stream going, and conditions are deteriorating by the minute and the fire was growing larger and larger. At this point it looked like the fire was going to jump the 15' alley and get into the Manchester Board Paper Company so everybody is running around frantically to try to get water on the fire to try to keep the exposure building from catching on fire. At this point, Captain Hinckley, (who throughout his career had sought out office jobs and was one of those full duty cowards that my buddy, Pete Lund used to talk about) came up and he was so excited he kept asking what railroad tracks the hose was lying over. It wasn't just us; there were several other engine companies that had by this point laid their lines over the tracks as well. He was so excited that if someone had grabbed him by the arm and said, "Captain Hinckley, what is your name, Captain Hinckley?" he would not be able to tell you. He had lost his head to that degree and was pretty much useless. Gene-O looked at him and laughed and the rest of us were chuckling at this guy because he was so panicked. Gene-O looked at him and told him, "Captain, it's that company that has all those big locomotives, yeah that's whose company it is, it's that train company." At this point, Captain Hinckley ran off to some other task that was seemingly important which to him was going to make all the difference in the world.

This fire taught me a pretty strong lesson. After we had set up the stang nozzle, Herman, Gene-O and Tupp pretty much disappeared, and made themselves scarce. By that time it was up to a 4-alarm fire and any time there is a fire of that magnitude you get all these off duty chiefs and assistant chiefs and fire prevention chiefs that come out of the woodwork and feel compelled to give orders. By order, I mean that they would tell you to run over here and get this, run over there and get that; a lot of which didn't amount to a hill of beans and it seemed to stroke their ego to give orders. About the time in the incident when we were washing the wall with a 1,000 gallon a minute squirt gun, Herman, Tupp and Gene-O had made themselves scarce so they wouldn't have to run these foolish errands for the self-important chiefs, the fire

Jake Rixner

chief of the city, Chief Lewis came by and looked at me and Captain Marsh and smiled and said that we were doing a nice job of washing that man's wall. There was an element of sarcasm in his voice but he did seem to be somewhat amused too because he could see the frustration in Captain Marsh's face and 20 Engine was coming in on the 3rd alarm and we had just radioed contact with 20 to piece into our line and pump into our line to elevate our nozzle pressure. At about that time, 20 ran over one of the couplings with their pumper, and ruptured our hose and consequently we totally lost pressure. This was pretty much my first major fire in Richmond and it was pretty much a wash.

For all we accomplished we at Engine Company 5 could have stayed in quarters and not much of the outcome would have changed. It was that embarrassing. Every fireman goes through this period of education where you learn to keep your head about yourself. This fire taught me the importance of planning and execution as far as water supply and to keep a cool head and know your limitations. For example, you can't lay out 1,000 feet of hose and try to use a 2" tip at 1,000 gallons per minute with one fire pumper producing pressure. It just doesn't work. The end result was that the Standard Paper Company burned to the ground before our very eyes. Another thing that was impressive during this fire was that there was a row of pretty good sized, factory windows that we were aiming for that faced the canal on the 2nd floor level from which the smoke conditions could be observed. When we first started trying to put water on this fire, it was heavy black smoke condition coming out of the windows. As the pressure increased, as the velocity of the volume of smoke really began pumping out of the windows, then within a minute or two it all went orange. All that black carbon coming at us combined with pressure and intensity of convection currents just pushing the smoke out of the windows; when it reached about 1200 degrees, the carbon ignited into a fire ball. This fire ball reached 25-30 feet out of the windows and even from the opposite bank of the canal; we had to back up from the intense heat that was put out.

Another thing that I will always remember about Standard Paper Company was that it was probably the only time in my 22 year career that I wish I would have had a camera because there were 3 pumpers; 17's pumper, 21's pumper and 6's engine all had their hard sleeves in the canal drafting and they were lined up in a one-two-three tandem fashion. It would have made for a neat picture; urban fire engines drafting from a canal. This saved a lot of city water and a lot of hose too. The canal was right there at the fire building.

These 2 ½ years of fires were instrumental in shaping my beliefs and outlook on how to manage a large, out of control incident. A lot of guys would lose their cool and run around shouting all kinds of things that if you kept your head sounded comical.

The next big fire I went to was at 15th and Cary on St. Patrick's Day 1985. There was a guy filling in at 5 Engine with us that day, Roger Neptune. Roger was from 13 Engine and had a reputation as a damn good fireman and although I had worked with him several times when I filled in at 13, I had never really been in a serious fire with him. Firemen call this "going down the hallway" and what they mean is that this is a guy you can trust your life with. He is a guy that has been tested under pressure in hazardous fire conditions where life and death are on the line. The metaphor of going down the hallway means that you never know what is at the other end of the hall; whether something is going to explode, or the fire is going to overtake you and it is dark, hot and smoky and it takes a lot of balls go to down the hall. Roger had the reputation of being a good fireman, but I had never been "down the hallway" with him.

On this fire we were first due on the 2nd alarm. It was a 3 story, 150' by 100' brick and joist tobacco warehouse. When we arrived on the scene, 6, 33 and 1 all had 2 ½" lines through the door and heavy black smoke was pouring from the roof and 3rd floor. There were no visible fire conditions from the street, but just heavy black smoke. Battalion Chief Griggs was already on the scene and a Truck Company had already placed a 35' ladder to the 3rd floor window and he instructed us to take a 2 ½" line up that ladder and get in the 3rd floor. I did wonder who had the 2nd floor covered because as we passed the oversized, large factory-type 2nd floor window on the ladder carrying our 2 ½" line, we could look in and see there was pretty good smoke and heat conditions on the 2nd floor.

The last thing you want is to be on a ladder 10' above conditions like that and have the glass fail and be enveloped in a bunch of fire. I was still a fairly new firefighter and a new firefighter for at least his first 3-5 years is still trying to get his feet under him as far as trying to figure out what the heck he is doing and that's where the other guys came in as far as teaching you the ropes.

Chief Griggs was a pretty good fire ground commander but did tend to get excited and yell and scream a little bit, but never would do anything to get anybody hurt. Now this fire went to 4 alarms and Roger Neptune ended up with the nozzle and was ahead of me on the line. When we got to the 3rd floor windows, Roger broke the window out and the smoke poured out so rapidly and thickly it was reminiscent of the fire at Standard Paper just before the carbon ignited. There was a tremendous amount of heat. The bulk of the fire in this building, we would later find out, was located in the open stairwell and a large industrial elevator shaft (12' by 14') had also become involved. In essence, there were 2 huge fires in what could be described structurally as 2 giant chimneys which terminated on the top floor of this building. All I knew at the time was that Roger had just broken out the window and we were trying to climb into the 3rd floor window into heavy heat and smoke conditions. In our favor, because it used to hold tobacco hogs (wooden crates of tobacco that were curing) the floor joists in this place were a true 2" by 12" dimensions, centered every 3 inches. There was little to no danger of the floor collapsing due to the close proximately of the floor joists that was used to hold tons of tobacco. The building was about 150 years old and did have old wooden floors that were impregnated with oil from the forklifts that used to move the tobacco hogs and it was ripe for burning. As Roger broke the window, I was on the ladder below him with my head at about the level of the back of his knees, (we were wearing the ¾ pull up boots) and I was steady hitting Roger on the back of his legs telling him to get in the window because I wanted to get into the building before the line was charged so we could get enough line in there and I wanted also to get in and see how much fire we had and thirdly I wanted to get the heck off the ladder in case fire did erupt out the 2nd floor windows. After I had hit him on the legs several times and he was not making any attempts to advance; from his waist up he was enveloped in smoke, so I couldn't see him from his waist up. He was a little bit angry with me and bent down to tell me it was too damn hot and as he was telling me this, I could see that his protective eye shield and flashlight both made of plastic were completely melted and distorted. It was indeed too hot for us to enter the building. He was angry with me because I kept hitting him on his legs telling him to get in the window.

   It wasn't long after we broke the window that this fire also got to the point where the smoke lit up. Chief Griggs noticed the density and velocity of the smoke and ordered us back down the ladder just in time that when we hit the ground, a fire ball about 25' shot out that window and showered everyone in the street with broken glass and debris. Shortly after that the remaining windows on the top floor went and this building was going away from here quick. The lesson here was that sometimes you do everything by the book, the 1st alarm company had the lines in at the seat of the fire attacking the stairway and elevator shaft, doing their best to get the stairway under control and the fire was just too big.

   The fuel, the fire load the amount of heat it was producing was just too big for the four 2 ½" lines that were deployed on it. This fire was the second time I got stuck on the stang nozzle that we had set up in the street. Cary Street between 14th and 15th Street is kind of narrow both sides of the street had three and four story buildings that were built about five feet off the curb line so you were literally in a small canyon in the street. 1 Truck set up their aerial platform there and 2 Truck had a ladder nozzle up and 5 Truck ended up in the rear with a ladder nozzle, and once we started to get some good water on this fire, the flame conditions blackened down, but it also caused the smoke to lose its buoyancy since it had cooled off to the point that it was not rising any more. It was coming down right to the street level and we could not see our hands in front of our faces out on the street. This was all well and good except for the fact that I was alone on the stang nozzle and anybody that didn't have a duty right in front of the building had left since the smoke had gotten so bad. There were other people around, I could hear them hacking and coughing, since we had taken off our self-contained breathing apparatus when we knew we weren't going back into the building. My SCBA was sitting in jump seat on 5's Engine at the hydrant on 15th Street on the alley between Cary and Main Streets. I entertained the idea of crawling up the street with the rest of the troops, except that each time I got off of the stang nozzle which is attached to a big metal platform that weighs about 80 pounds, the gun began to move because of the water pressure being supplied to the nozzle. If I had left it, God only knows where it would have taken off to. So there I was shooting water into a 2nd and 3rd floor window into a building that I could barely see, then the smoke came down and by this point I couldn't see at all, so I had to listen for a hollow sound that was water going through the window as opposed to the water hitting the wall or washing the bricks. Remember, I was not the only player here; there were about 4 or 5 other fire streams in that block and I could hear that some knucklehead had moved his stream to the point where it was washing the brick wall. The problem with this was that I had a 3-story building towering over me that some clown is washing the mortar out from between the bricks, I can't leave the nozzle and I don't have a shut-off so it is only a matter of time before this guy washes out enough mortar that he is going to cause this wall to collapse into Cary Street. Thank goodness, one of the chiefs heard the same thing I was hearing and I soon heard someone getting their butt chewed out about 50' east of me. I was laying down on my hands and knees with my face down on the granite curbing trying to suck up what little air I could from the street, which in addition to experiencing this type of smoke conditions outside of a burning building, was something I only had seen a couple of times before in my career. Big fires can teach you a lot if you take the time to study what happened and try to educate yourself. I recommend that you use a critical analysis to reconstruct what happened at the fire and make it a learning experience.

There will be plenty of mistakes, plenty of people that you can learn from and that can teach you. There will be seasoned veterans who make mistakes and those guys who are quiet in the firehouse that become shining stars. The most obscure men sometimes step up and become heroes in some incredible ways.

We spent about 4 or 5 hours at this 4-alarm fire on Cary Street and unfortunately as most big fires go, it is now a parking lot today due to the amount of fire, the age and condition of the building and the amount of fuel inside the building. When you get a good fire going, it is hard stopping it, short of channeling a major river through it, it is going to burn the amount of fuel until it is down to your level.

Just four days later, there was a fire just up the street, but I was off for this fire. This was also the shortest employment on a part-time job that I ever had. I had been hired by Miller and Rhodes, a major department store in Richmond, as a truck driver based out of the main warehouse off of Hermitage Road. It was my first day on the job (I was being trained) and I had just completed all my paperwork when I went to get lunch at noon. I looked up and could see a large column of smoke rising over the downtown area. I stopped in at the closest fast food joint which was only a couple of blocks from Engine 5, for a couple of burgers and a soda and knew that with that much smoke, 5 would not be in quarters, so I thought I would just go and look at the teletype and see what was going on. I pulled in the alley behind 5's firehouse and used my door key to let myself in the building (which was kept locked since it was located in a pretty rough neighborhood). I walked up to the watch desk to check the teletype and I could hear a diesel engine sitting on the front ramp, just idling. I opened the blinds and peeked out the front window and 19's wagon was on the ramp and 19's engine was parked against the curb across the street. They had been sent from the far west end (19 is a pretty slow engine company) to transfer to #5's quarters to protect 5's first due area while they were at the big fire. This fire was located on a little street called Shockhoe Slip, just off of Cary Street, between 14th and 13th and Virginia Streets, in an area that was beginning to be renovated into the quaint shops that are there today. A fire on a weekday, at 11:00 in the morning had taken off in a 4-story brick and joist building that housed Southern Railway and Supply Company. This building was at least 100 years old and when Engine 6 arrived, they reported fire conditions on all 4 floors of the building.

There were also about a dozen cars burning that were parked in the lot just adjacent to the fire building. Because of the geography of the building, the fire was spreading in 7 different directions and fires in buildings on 4 different streets. I unlocked the door and let 19 back their equipment into the firehouse. Someone from 19 had asked me if I was coming in on the recall. I asked them what they meant. They told me that the fire chief had put out an order that all off duty firemen are to return to their stations. I told them that I was working a part time job and hadn't heard about the recall. I called Fire Communications to find out where we were supposed to muster. There were 5 firehouses in Richmond that had reserve pumpers and there were 2 other houses that had spare ladder trucks. These pieces of apparatus were used when a station's regular piece was in the shop. The typical plan for the off duty men on a recall was for them to staff all the reserve engines that we could. Richmond had two foam trucks and other equipment that could be pressed into action if needed. When I contacted Communications, I was told that there was indeed a recall and everybody from my station was to report to #1 to staff reserve Engine 1. I quickly got my turnout gear and headed over to #1, thinking that I would be making overtime today and could possibly be going to a big fire (the guys from 19 had told me that this fire was already up to 7 alarms). Richmond had 30 fire companies at the time, and there were only two or three companies available to run calls for the rest of the city.

When I arrived at #1, there were a number of guys already there and 1's reserve engine was a 1962 Oren with a gasoline engine and a 1250 pump. It was a cab forward model with jump seats and with 5 of us on the back step; we had about 13 men stuffed up on that thing that day. We had a pretty well staffed engine company, the problem was that the reserve engines at that time only carried hose; no nozzles, no air packs, no tools. All the ancillary equipment had been removed. When a truck went to the shop and you had to get a reserve piece, you ended up pretty much stripping your piece to equip the reserve piece. It was a pain in the neck to get a reserve piece since it would take two hours to transfer all the equipment that was needed. There were even reserve pieces that didn't have any hose, so the men would have to switch the hose over to the reserve piece as well. The problem during this fire emergency was although we had inch-and-a-half hose for fire attack and 2 ½" for fire attack and water supply, we had no nozzles, and not even a hydrant wrench to get water. That set off a quest. We went into the supply room of #1 and quickly found a hydrant wrench and some other smaller tools. I remember that we couldn't find an inch-and-a-half nozzle, so the plan was to head over to #11 since they had the hose repair shop there, and it was known that in the basement was a lot of nozzles, extinguishers and pike poles as well as other tools that might be useful to us. We were hoping the whole way to 11 that we wouldn't be called to a fire; since all we had was hose and a hydrant wrench. I was thinking on the way over as I rode the back step, that it would be our luck to get called to a fire without tools or air packs and we would end up in a 2nd floor hallway with an inch-and-a-half line using my thumb as the nozzle. It was times like that that we tried to keep the mood light and find humor in the absurd which was a good way to keep your sanity in that type of situation. We got to 11 and there was an engine company from Henrico County (a neighboring county to Richmond) which was very rare.

The city believe the mantra that they were self sufficient and didn't need assistance from surrounding areas, where the counties were much more open to the idea and called Richmond several times to assist them with their big fires. The "old timers" in Richmond believed that it was insulting to have to call for a county to help. Upon our arrival to 11, one of the off duty battalion chiefs (who was of this mindset) showed up and the very first thing he did before checking with the officer of the reserve engine to find out that we didn't have any equipment, was to merrily dismiss Henrico to go back to the county. The way he said it made him come across as a pompous ass because there are only two or three engines left available in the city and this county is nice enough to send an engine to one of the most fire prone areas of Richmond when the city was in need. In essence, he sent home a well-equipped modern piece of equipment in exchange for a 23 year old fire engine with no equipment on board all in the name of foolish pride. This was another irony of ego that I have seen throughout my career. As things turned out, we were never summoned to the fire ground, and as I recall, there was fire in 13 different buildings facing 4 different streets and it was brought under control by sheer bull work. In fact the C-shift at #5 was on duty that day and they actively stretched a 2 ½" line up a rear staircase of a 3-story brick building and put out fires on all three levels and essentially kept the building from suffering extensive damage even though there was fire on all three floors.

After a tremendous effort by the 5 men on C shift, the building owner congratulated them and proceeded to inform them that he was glad they were able to save his building because he had just leased it out and was getting ready to move to the county. Wilbur replied to this by saying, "Damn, I wish you would have told us this two hours ago we'd have let this son-of-a-bitch burn!"

Once we had acquired enough tools and equipment, we returned to #1 and sat there and another battalion chief came by and by the luck of Richmond, Engine Company 23 was the only company available to cover the city. There were no hook and ladder companies and the 19 engine companies were on the scene with the exception of the reserve engines and ladder trucks that the off-duty guys had been able to put in service due to the recall. Chief Tyler came by after the fire was over and we were looking through the teletype at the run down of all the units and Chief Tyler pointed to 8 which was a house of older guys, most of which had about 30 years on the job. This was a slower house where the older guys gravitated toward to wind out their time before retirement at age 60.

As Chief Tyler pointed to the teletype at the last engine sent to the fire and said, "See right there, 8 put the fire out." That was a lighthearted moment, and we all knew that it wasn't true. Back then in the fire bureau, the older guys were highly respected by the younger men since they had put in their time and had all served in a day where there were no air packs and they hung on the sides of the hook and ladders and stuff like that. The amount of respect for the older guys was phenomenal and well deserved. Places like 19 and 8 were saving graces for the older guys who didn't want to be working the fast pace of 11, 5, 6 or 1 which were the busy stations during the 1980's.

The next big fire was at 3100 Hopkins Road. This fire really stuck with me for a long time. It was a huge 1-story brick building that looked like it was some type of warehouse for some type of railroad operations at one time. It had railroad tracks on both sides and there were several buildings all one story that were about 120' wide and at least 800-1000' long, and were in rows like army barracks. The reason this fire stuck with me was that we went on the 3rd alarm and I was on the back step and when we came in on the 3rd, Chief Griggs was on the scene, barking out orders and commands in staccato fashion and what we had was 120'wide by 850' long building that was fully involved. The trees near the building were fully involved and there was a similar building just 100 feet away to the rear that was smoking and had several large propane cylinders that were beginning to discolor from the heat. In hindsight and using a critical eye, what was amazing to us as we got to the scene was that the main exposure building was to the rear and it was 1000' long with two large propane cylinders that were ready to explode, and all the first two alarm companies were stationed on the front of the building where the only exposure was a row of trees. When we arrived on the scene the entire building front to back, top to bottom was a sea of orange.

The roof had caved in over about a third of the building. There was no saving anything in this building. It was gone. Our officer had the

Jake Rixner

presence of mind to have us lay out from 10's engine between the two buildings to get some water on the propane tanks before they exploded and try to keep the only exposure of any value from becoming involved in this huge fire. Chief Griggs ended up striking 6 alarms on that fire and now that I was maturing as a member of the fire service, I remember thinking that he could strike all the alarms he wanted to, but everything that is on fire is gone, and he needed to help keep the second building from catching on fire or it was going to be gone as well.

Looking back, big fires, in my opinion were not a whole lot of fun but you can learn a lot of strategies and tactics at these fires, but the most rewarding fires for most city firemen is the one or two room fires where you really take the grit and guts and determination to bring under control and you are making a difference in someone's life by saving their property. In most of the big fires, everything that you encounter when you get there is either destroyed or in the process of being destroyed and it is mostly a defensive attack rather than an offensive attack. Most big fires are intense for about the first 15-20 minutes, some lasted for an hour, but pretty soon you find yourself babysitting a large water cannon and most of the time, freezing your butt off and there is little there to be intellectually stimulating or challenging.

- Chapter 10 -

Refugee Truck Companies

We were back to a nice full, crowded firehouse and each company had different personalities, we were a very diverse group. Engines and Trucks when they run together are notorious for teasing each other over all kinds of stuff. Especially 6 Truck, even though they were further away from their area, they were famous for being slow getting out of the house.

Lieutenant Nappy was taking what seemed like forever getting out because they were running from 5's quarters instead of 14's quarters, still covering their original area which was further away. Lieutenant nappy who was about 60 years old would saunter over to the mid 70's model Seagrave tiller truck which was not that tall of a truck; in fact it was kind of squatty, but it would take him forever to get into the front seat.

Captain Emerson worked on the same shift as Nappy and Emerson made the comment to me one time for me to "help that old man up into the truck the next run they get; they are so damn slow getting out of here, they are going to get cancelled from their call before they even get off the ramp." I knew Emerson was only messing with Nappy because no matter how old a man gets he doesn't want anyone to think he needs help getting up into the truck. I had a pretty good sense of humor so the next run they got, I positioned myself at the officer's door and Nappy came shuffling out of the bedroom in his slippers and proceeds to put one foot up on the running board to get into the truck and at this point I grabbed him up under his arm as if to help him, and this 60 year old man in about a nanosecond turned around and slugged me in the stomach so hard it knocked the wind out of me. I never thought that an old guy like Nappy could punch that hard, but he took my wind away and started laughing. As they pulled away, I was left standing there in the middle of the apparatus floor trying to catch my breath.

During my first 2 years at 5 Engine, 6 Truck was temporarily stationed with us while they were getting their new quarters built at the same location as their former house. As soon as they built the new 14 Engine House, 6 Truck moved out and we enjoyed about 3 weeks of being a single company house. There was now almost a television for each man, easy access to the phone, just little things that were nicer since there were half as many people in the house. When there were 10-12 people on the shift we had to go by the majority rules to decide what to watch on TV, etc. We paid a house tax (or house fund) of $7.00 per pay which went into a kitty that paid for a separate phone line, cable TV bill, new televisions, coffee, and other household sundries. It gave us a lot more freedom as far as access to a telephone or television as a single house. It was also more conducive to go into the bedroom and read a book or study since it wasn't so crowded.

On the down side, it was kind of fun like having 11 brothers living with you and we had some great games of dominos, cards, ping pong and pool. We had one lieutenant, Lieutenant Rove who used to love to play board games like Parcheesi, chess, and he was good.

We were only by ourselves for about 3 weeks when 3 Truck was backing into quarters one night and they tried to use their 1971 American LaFrance tiller truck as a battering ram and tore half of the front wall of the firehouse down. I wasn't there so all I know is they were backing into the firehouse and the driver put it in reverse and just floored it trying to show his ass a little bit and the tillerman wasn't ready and the trailer struck the front wall of the firehouse and knocked about two tons of brick out of the front wall. It was around Christmas time 1983 when 3 Truck moved in with us. It took them a good 2 months for a masonry contractor to fix the firehouse, so Captain Emerson dubbed our engine house as the "house of refugee truck companies" and the name kind of stuck and we used to give 3 Truck hell about this.

Firemen are famous for starting and spreading rumors and keeping stuff stirred up. The city budget was always a hot topic as discussions would come up that they were going to lay off firefighters, cut 10% of this, 10% of that and most of the time there was nothing grounded in reality.

What happened in April 1984 surprised everybody. City Council did cut the budget. 3 Engine had been out of service since June 1981 leaving 41 in the house at 6th and Marshall alone since that time. It is not unusual for a truck company to be in a house by itself; in fact the firehouse that is in the movie Ghostbusters is an actual truck company #8 in lower Manhattan in New York City and still in service today. We were notified that in about a week, #3 Engine House (which housed 41) was going to be closed and 1 Truck was going to permanently move to #5. We were no longer the home to refugee trucks, now we had an orphan moving in with us. This made for some out of the ordinary times.

April 1984 was a very exciting month for me personally because that was the month that my son Jimmy was born and my wife and I also bought and moved into a brand new house that we had built for us in Mechanicsville. At the firehouse we were out of service for 2 days and used city school buses to move the furniture and supplies for the service truck (which were in the basement of Engine 3's house) and that was the bulk of the work there, making trips up and down those stairs into that dungeon and moving all those supplies over to #5. These were three major events in my life; the firehouse life changed, my home life changed for the better, moving into a brand new house and a new baby at home.

When 1 Truck moved in we knew it was permanent and we had to get used to the new situation. Donny Viar, Pee Wee, Eddie P. and I guy I'll call Hal made up the shift at 1 Truck. The city had recently taken away 1 Truck's tiller truck and made it a reserve truck and bought a brand new Firebird which was a 100' elevated articulating type of platform truck and I think at the time it was the only truck in the fleet that was not a tiller truck. Pee Wee had just come to work at that time and Viar and all those guys and it was neat within a couple of years with guys transferring in and out because they had a straight truck, these guys didn't know how to till.

Lo and behold one morning we showed up to work and the Firebird was in the shop to be worked on and Captain Holt was the truck captain and at line up he told us the truck was out of service and they would be using a reserve truck which was an old hook and ladder truck, in fact, it was old #4's truck which was a sister truck (two rigs bought at the same time) to #6 truck in the mid-70's. So Holt told us that someone had to go work at 13 because they were sending us a tillerman to till the reserve piece. He asked if any of us new how to till and I raised my hand because I had learned to till when I was in West Lanham, and he was kind of miffed and asked why I hadn't told him and I told him I was assigned to the engine company and besides that engine men don't have to know how to till. Holt took me aside and asked me again if I was serious when I said that I knew how to till and I told him I had known how to till since I was 18 years old. Holt put the truck back in service and made arrangements with his counterpart in the engine company, Lieutenant Val Jean, and they made a manpower swap; Pee Wee was to run on the engine that day and I was assigned to till 1 Truck.

After this, Holt said that he wanted to use one of the reserve tiller trucks and practice tilling regularly and get everyone in Truck 1 to be checked out to till. Right after clean up, we went out and rode around for about an hour. When we backed into the firehouse, Holt pretty much gave me his blessing and told me I was now officially qualified as a tillerman. This made me chuckle to myself because that really wasn't the exact procedure to become a tillerman. Usually a safety officer had to come out and administer the driving test and then you would be issued a paper driver's license saying you were qualified to till. Every time there was a reserve piece from that time onward, I would till for them, since he didn't have time to train the other men to till. That set me up for one of the most remarkable rescues I have ever participated in. There was a guy, Nick, who had a little more time on the job but had just been transferred to 1 Truck, so he was new to the house, and he was riding the side of the ladder truck one night I was tilling. The engine company was at 2nd and Jackson on a car fire about 9:00 p.m.

It was a warm evening and we had the door up and about 5 minutes after the engine went on the call, a car pulled up onto the ramp and a gentleman who was very excited came running up to us and said that there was a fire at the 700 block of W. Leigh Street. We fired up the hook and ladder and could see the fire from the ramp, as the 700 block was only 3 blocks from the fire house. There was fire out the back on the first floor of a huge 100 year old frame 2-story house.

There was so much fire coming out of the house that the alley was lit up as though the sun was rising behind this joint. We pulled up and called for the assignment which ended up being 10, 6, and the 2nd battalion chief. The hook and ladder trucks don't carry any water or hoses, just tools and ladders. The front of this house was boarded up so I got out of the tiller seat, grabbed an ax, and began prying the plywood off of the front door. Nick came to help me get the board off and we go into the first floor. There was still no engine company on the scene. It would be about 2 more minutes before 10 would arrive so there we were, going into a burning building with no water. The entire kitchen and dining area are fully involved in a massive body of fire. Nick and I proceeded to search the first floor back away from the fire, still thinking that this was an abandoned, unoccupied building since it was boarded up. The staircase was inside the front door beyond the grand foyer that was characteristic of the old Victorian style houses of that time period. After we searched the first floor, we scooted up to the second floor.

Because the fire was so intense and aggressive, I was going to just take a peek around, an in and out type of thing, because there wasn't a lot of time until someone got water on the fire. The last thing we wanted to happen was for us to be up there groping around and get cut off or trapped by the fire. When we got to the second floor there were four doors in the hallway.

The very first door I came to was padlocked from the inside, which indicated that there was someone inside the room. I kicked the door in and Nick continued to scoot toward the front of the house. I had to kick the door three times before I could get in. As soon as I got the door opened, I heard someone choking and coughing in that room. At that time a thought ran through my mind before I entered the room that supposed whoever is in here has a gun; he is going to think someone is breaking into his room and I'm liable to get shot right here. With that thought in mind, I figured my best chance to bum rush the sound of coughing. I couldn't see him, but I charged him like a bull and the man was sound asleep and when he heard his door being broken in, he sat up in bed and then was breathing the smoke and began coughing. I tackled this guy in his bed and by this time, he didn't know what was happening so I told him, "fire department, we gotta get going, let's get out of here, the house is on fire." So I snatched him up but when I got him back to the hallway, the fire had come across the first floor ceiling and had started to lap up the well hole of the stairway which told me that we were on borrowed time and had to get out of here now. As soon as the man saw the fire, he began to panic and broke away from my grip and headed for his room. At this point I could hear Nick wrestling with someone else and I heard a window break in the room that Nick was in. My guy sort of came to his senses, and he was a bigger guy and I wasn't sure that I could have physically removed him with him fighting me, but I told him we could make it and we had to leave now and once I got him halfway down the steps, he took off like a scared rabbit. I got him outside and Lt. Bridges was out there and the victim was exhausted and the smoke was taking effect. They called for an ambulance.

About that time, I saw Nick on the front porch roof with two other fellows. What had happened was when Nick kicked another door in, he heard window glass break, but remember, the front windows were boarded up, this guy was asleep and when he heard his door being broken in, he instinctively reached up with his arm and put it through the front window and cut himself pretty good. He was bleeding and had cut an artery and the blood was just squirting. There was another fellow in that room sleeping and Nick cleaned out the window with his tool and knocked the plywood off the building and got both of them out onto the porch roof. By the time Nick cleaned the window out and called for a ladder to be raised, he went back to the doorway to the hallway and could see the fire coming up the steps. He closed the door to buy them some time and got the two men out. In all this excitement and confusion, 10 and 6 show up and put water on the fire and put it out. My smoke inhalation guy was the last to be transported because the arterial bleed victim went first.

Lt. Bridges came up to me and asked what room I got my victim out of and I told him the first door on the left. Bridges grabbed the police officer and told me to take the cop up and show him where he was. The guy had told the police officer that he had a 357 magnum under his pillow and he wanted the policeman to retrieve the gun so it didn't get stolen. My guardian angel was working overtime once again, by my deciding to rush the guy so I wouldn't get shot. I have been to hundreds and hundreds of building fires and that thought had never entered my mind, but for some reason, that night it did and I may have been very close to getting shot if the guy had another second or two he may have had time to react to what he thought was an intruder. Firemen don't sit around the kitchen table and talk about stuff like this, but in the venue of a book I felt that it was appropriate to mention.

There are lessons here for young firemen to remember. First, that just because a building looks vacant, doesn't mean that it is vacant. Secondly, you can do a lot of good and conduct a pretty aggressive search without the presence of a water hose or engine company on the scene; you just have to be damned careful.

The City had just purchased three new Seagrave hook and ladders (tractor trailers) and these trucks were beautiful, state of the art but nothing fancy, they were good solid hook and ladders. Richmond was big on wooden ladders and there was one Fire Bureau employee whose sole job was to repair the ladders and make any cabinetry that was needed at the firehouse. The wooden ladders were kind of rare on the east coast and more popular on the west coast of the U.S. It was my understanding that these three new hook and ladder trucks had the last wooden Duo Safety ground ladders ever made. Duo Safety wasn't selling enough of them so they were getting out of the wooden ladder business.

Our fire chief was pretty sharp and had a couple good reasons for liking and using the wooden ladders. One reason was that they conducted electricity much less than other ladders. It cannot be said that they didn't conduct electricity at all, because any moisture on the ladder itself could get you shocked if you threw the ladder into a power line. We still had to be careful, but these ladders were a whole lot safer than the aluminum ladders which would definitely light you up. The second reason he liked the wooden ladders was that we had six men assigned on a hook and ladder truck, a driver, an officer, a tillerman, and two or three men in the jump seats depending on manpower. A 35' wooden ground ladder is a heavy duty ladder that takes three good men to lift it and four men can throw it pretty easily. If you don't know how to throw a ladder, a 2-section, 35' wooden ladder will definitely make a fool out of you. The chief could justify the manpower based on the fact that 3-4 men were needed to throw this ladder.

So as 1984 came to a close, Ladder Company 1 was now permanently housed with 5 Engine in the little yellow brick firehouse on Leigh Street.

- Chapter 11 -

New Blood

The winters were kicking everyone's ass. Herman and Tupp were both in their 50's and with Engines 3 and 31 gone, and the Jackson Ward neighborhood deteriorating, it was obvious that 5 would only get busier and busier. Herman Brawn was the first to put in transfer papers. The city had 21 firehouses, of these about 13 saw fire on a regular basis. The outlying stations had more stable neighborhoods and since we didn't run mutual aid with the surrounding counties there response area was limited. Stations like 8-16-19-23-24-25 were almost like retirement homes when compared with 5-6-1-11-13-10-12. Since the pace was slower, and because these guys had given their best years in city service, they were rewarded with assignment there. I was glad there were men who wanted to go to the slow houses, because if I had to work there when I was in my 20's it surely would have driven me nuts. It took him a while to get the transfer and when Herman transferred to 8 we got Grayson Finner assigned from the A shift or the "A TEAM." As we called them. It wasn't long before Tupp was pulling strings trying to get to 8. He got there by convincing the chief to transfer Steve Willy to 5 from 8. Steve had been assigned to 8 for 12 years having gone there straight from drill school. He didn't want to go to 8 at first but quickly became used to the slow pace. Sending him to us was like the ultimate insult to Steve…talk about culture shock.

Every run after bed time you could here Steve bitch and moan, he was used to sleeping all night, and Jackson Ward didn't really come

Jake Rixner

alive until about 9 pm. We used to tease him and try to help him make the transition, but it was clear he was unhappy with his new home.

We were returning from a run one night about two in the morning, I was driving the engine and Steve was riding with me. The wagon was northbound on Jefferson St and when we passed Grace St. something caught my peripheral vision. The Farm Bureau building on my left was only 6' from the narrow street and had an indention in the brick wall for a fire escape door. Backed up against the door were two men who were having sex with each other. I locked the maxi-brake down as the big Mack pumper slid to a stop while laying on the air horn to alert the wagon. The homosexuals were really going at it and the sliding fire engine and air horn ruined their moment of paradise. I remember being less than ten feet from them and pointing out the window and shouting, "HEY, YOU'RE FUCKING IN THE ASS!" Lt. Pulliam came running back to see what was wrong just as these two lovers dashed from the doorway pulling up their pants as they ran west on Grace. That was about the last straw for Steve. He looked at me and said " Rixner you're a fucking nut case and your going to get us killed someday" I remember finding this incredibly funny and laughing at him for a good five minutes. His father-in-law had a kitchen cabinet business and wanted to retire. Steve left the Bureau and took over the business. We were sorry to see him go.

I was living in a nice apartment on Chamberlayne Ave. I had always thought it would be nice to have my own place after living at the firehouse in Maryland, but it was very lonely coming home to that empty apartment. My girlfriend, Julie, was in her second year of college and didn't want to go back; she had been hinting about marriage for some time and now with a steady job and benefits I figured it was time to settle down. I was shopping for a diamond ring, but the prices seemed way too high. Herman Brawn gave me the number of a store in the West End that he dealt with and even called the guy and told him to help me out. I went out to the store and the guy had several nice rings at decent prices. He seemed a little too nosey for me and when he asked me who the lucky girl was I told him I was marrying Herman's daughter. I have always enjoyed a good joke but the look on his face was priceless. Here was this nice young white boy marrying a black girl. I found that in many ways, Richmond was way behind the times socially. Some of the folks were still fighting the civil war. In fairness Herman had several nice looking girls, and he enjoyed the laugh as much as I did. We set a date of April 23, 1983. It was the closest to a warm weather vacation a rookie had a chance to get.

I had read a number of books over the years about the New York City Fire Department and wanted to go up and ride with them before I got married. I had 9 days off before the wedding and figured I'd better do it now, and get it out of my system. I packed up and drove to New York on April 14th. I didn't know a soul in New York and when I got into the city I rode around until I found a firehouse. As it turned out I ended up knocking on the kitchen door of Engine 234 & Ladder 123, and the 38th Battalion Chief in Brooklyn. It was a rough neighborhood and the fireman asked to see my i.d. thru the glass window. I showed it to him and he let me in. I told him about looking to do a ride along with someone and that's all it took. There was an older Lt. in 123 truck (whose name I wish I'd written down) that took me under his wing for the night. We went to two fires and a couple of other runs and he shared his thoughts about the job. He had over 20 years on the job and was getting ready to retire soon. He showed me their training magazine W.N.Y.F. It listed the top 25 engines and trucks as far as runs. 75 Engine in the Bronx was on top of the list and he had a friend there. He called 75 for me and set me up to ride there the next day.

The next morning I left 234 & 123 and rode around the city sight seeing. I went to the famous Engine 82. <u>Report from Engine 82,</u> was a best selling book in the early 1970's about the F.D.N.Y. and the crumbling neighborhoods in the South Bronx. It is always neat to go to a place you have read about just to see if it looks like you had pictured it in your mind. 82 Engine was exactly as I'd pictured it with the exception that it was filthy dirty.

75 Engine was a real treat. Their firehouse was exactly like 10 Engine in Richmond. The house contained 75, 33 Truck and the 19th Battalion Chief. 75 went out about every 15 minutes. I was riding in the front seat between the driver and officer. The radio talked so fast it was hard to understand what they were saying, but I remember the officer telling the dispatcher "Engine seven five to the Bronx, 10-84, looks like a job K." several times that night. We went to 5 working fires and these guys didn't quit. The other thing that impressed me was no one yelled or got excited. They were very professional. The next morning I left New York and headed home, although I'd planned to spend 9 days there, 2 was enough. I had also planned to pick up an application, but 5 Engine in Richmond was busy enough for me.

Back in Richmond I stopped by the firehouse to pick up my paycheck, Captain Butler was acting Battalion Chief and was leaving as I came in the door. "Hey Captain, you got any overtime available?" I asked. He never stopped walking and didn't even look my way and said, "21 next day." I wasn't sure if he was serious, but I knew he had a very dry sense of humor. The next workday I reported to 21, I didn't know if I would be working or not. As it turned out he was serious and it was a nice chance to make some extra money for the honeymoon. Butler came by that day making his rounds and when he left he stopped as he was getting in the Battalion car and came back in and said, "You sure you want to do it?" Meaning are you sure you want to get married. "Yeh, I'm sure" I replied. Butler just shook his head and left.

About an hour later the box opened up for a fire at Phillip Morris cigarette factory on Commerce Road. We were first due Engine as the assignment was 21-13-17-22-45-48-53. I was standing in the jump seat across the motor box from a guy from 21 named Jim. Jim was telling me that Phillip Morris had a fire brigade and we had nothing to worry about. As we approached the security gate on Bells Road I could see a guard waving us into the plant and appeared very animated. We stopped near the sprinkler connection and the Lieutenant ordered us to take the high rise bag into the building. Jim and I went into the building that covered acres and acres of ground and soon were climbing some open metal stairs in moderate smoke.

Now, I have never been a cigarette smoker but the tobacco smoke was everywhere. If you were a smoker this had to be a treat, but for me it was miserable. The building was so big and the smoke borderline so that you didn't want to put your face piece on and use you air until you absolutely had to. The air you saved just might get you back out of the building and save your life if things should suddenly go wrong. Fires in large buildings like this have a long history of suddenly going wrong quickly. We kept climbing stairs and soon we exited a door onto the roof of one of 8 towers that rise another 80' or so above the factory floor. Both the Lt. and Jim suddenly began to panic as we were now above the fire. I remember going over to the edge of the roof and looking down.

This tower was set back from the outside walls of the rest of the 25' high building. 5 Truck was coming in through the Commerce Rd. gate and I could see that due to our height and location there was no way their 100' aerial was going to reach us I remember thinking that Julie is really going to be pissed at me if I get myself killed less than a week before our wedding. I gathered up my hose and started thru the door to go back down the stairs. The Lt. Looks at me as if I'm crazy and asks "Where do you think you're going?" "Well Lieutenant," I said, "since we are now above the fire by way of you astute leadership, with no other possible way of escape, I will now be descending these stairs and by the grace of God I am going below the fire level and try this whole firefighting thing again". It broke the ice and they followed me down the steps. To this day I think they were prepared to sit on that roof and pray for somebody put the fire out. As we went down the steps about 30 feet above ground level we meet Captain Butler and the men of 13 Engine who had found a duct fire. It seems they used duct work to deliver the tobacco to the cigarette making machines. Butler looks at me as we come DOWN the steps as if to say "where have you-all been?" No words are needed as he knows these players better than I do.

We tear open the metal ducts and chase fire for about an hour before Butler puts it under control. For the next two weeks my chest burns from the effects of the tobacco smoke. To this day I hate the taste of cigarette smoke.

Julie and I were married on April 23, 1983 and she planned and pulled off a beautiful wedding. All our family and friends were there. It was a wonderful day. In fact she was having so much fun at the reception; I had to drag her out of there. We honeymooned in Florida and went to Disneyland and all the other sights including Daytona Beach. Being married required some major adjustments in the way I was living, but it was nice to have someone to come home to. One of the big adjustments was giving up the weekly trips to Prince George's County, Maryland, and all the fires there.

The honeymoon was wonderful and it was a nice break from smoke, sirens, dirt, and danger, but after two weeks, I was itching to get back to the Firehouse.

The evening of June 9th the box opened for a house fire at 1913 Rose Ave. Rose Ave was in lower Barton Heights. We were second due. The area was full of 5,000 Square foot, wood framed, Victorian style homes. We didn't run up there much, but when we did it was always a kick-ass fire. We arrived first and one of these old mansions had fire 25' out the windows on the left side, second floor. I took the pipe and Duke was right behind me. When we got to the top of the steps there were two rooms freely burning to our left.

A Lieutenant, who I'll fictitiously call Jones, was running around in the foyer downstairs and was yelling at us not to go in there. I knocked the first room down from the hallway but couldn't get the second room, and we were just playing with the tongues of the flames. I looked at Duke and he looked at me. "What do we do?" I asked. Duke said, "Let's go." So we advanced forward and knocked the second room. On the way across the first burned out room, an ember went down my ¾ boots. I had the water aimed at the fire coming through the door and shook my leg; the ember went further down and lodged behind my knee. The pain went from bad to worse and I shut off the pipe and stuck it down my boot and filled it with water. When I shut off the water the fire rolled directly over us and Duke asked me "What are you doing?" "Sorry Duke, but I had to take care of that." After the fire the loo was upset at us. We tried to explain what had happened, but he didn't want to hear it. I finally heard enough and told him; "If you would have been where you were supposed to be instead of running around screaming like a little bitch, you'd know what happened." It was a pretty cocky thing to say, especially for a kid 5 days shy of getting off probation. The next morning this lieutenant woke me up and asked me to come to the office. I figured he was going to apologize and when I get there he's already got the captain there and proceeds to tell his story. The captain takes a long drag from his cigarette and looks at him and says, "You know Lieutenant, we've got a lot of problems in this company, but being too aggressive isn't one of them." Man I wanted to kiss the captain. The loo sees he's not going to win, but it will be a long time before I can respect him, and we wouldn't get along for a long time.

A couple of days later I'm driving the wagon when we are sent to Baker St. in the Gilpin Court. Projects. We pull in to see heavy black smoke pouring from the 3rd floor windows. As a courtesy, I look at this lt. and ask, "Which line do you want?" He's got his handkerchief out and is wiping the sweat from his forehead, his eyes are fixed on the smoke and it looks like his little world is coming to an end. I go to the back of the wagon and tell Grayson, "Look at this wizard; I think he just peed in his pants." We take the line with Grayson on the pipe and go up and knock down two rooms of fire.

We never saw our boss the entire time we were in the building. When we get back to quarters our boss wants to do some team building bullshit he learned in college. Grayson, Gene-O and I let him have it with both barrels. Grayson said; "Look here college boy, you can talk on your radio and throw books at the fire all day long, but until you grow a set of balls, the fire isn't going out!" "Man, what did we do to be blessed with this asshole?" I asked Grayson right in front of him. He slams a book on the table and storms out of the room. Oh, well, so much for team building.

The mood was pretty somber the next couple of days and our coward boss took about a month off to take care of personal business. Gene-O did some checking and reported that our boss was in a rehab center for a drinking problem. I didn't care where he was I just wanted to get back to the fun days. Duke was promoted to Lieutenant about this time, and we got a rookie by the name of Bryan Lam. Captain Butler was sent to 49 and Captain Richard Emerson went to 41. Richard had 23 years in the Bureau and had never been in a "Hook & Ladder Company" as he called it. And we could tell by his attitude, he wasn't interested in learning Truck work now! We got a young Black Captain named Marsh. From number 12 Engine. Engine 12 was a good engine that went to lots of fires but no one seemed to know this Captain Marsh. We quickly got on the phones as each man on the shift was looking for information on what kind of guy he was. In just two short years the entire shift had changed except for me and Gene-O, but one thing never changed, the fires kept coming in day and night.

Captain Emerson was now down at 41 and was acting Battalion Chief 53 when he called me at home one morning, "Hey kid, do you want to work overtime today?" "Yes, but we work tomorrow," I answered. It was against the rules to work over 36 hours straight. "I know but it's the 4th of July weekend and I've called everybody on the list." I reported to 5 with the "A" shift to start my 48 hr. tour of duty. July 3rd was an extremely hot and muggy day even for Richmond.

We had the normal 4 or 5 runs during the day and everyone was in bed when the box opened at 2358 hrs. for smoke in the area of 1st & Grace Sts. Little did we know, it would be the last time we got to sleep that night. The assignment was for 6-5-10-33-41-43-53. We met 6 at 1st and Grace and something was definitely burning in the area. The two hose wagons stopped side by side and the officers discussed which way to go. 6 ended up going south on 1st while we went west on Grace. Smoke was laying low in the streets and you could see the headlight beams stabbing through the smoky fog. The smell was that old familiar plaster and lath smell of a working building fire. One block up we found the fire at Foushee and Grace. The Capri Restaurant was a 2 story brick and joist structure about 30x75' Smoke seemed to pour from every crack and crevice in the building. Henry was acting lt. and radioed the correct location for the fire so the other units would come to help us. A fire hydrant was cattycorner from the building and our engine pulled up to it and put in. We laid our hose with the wagon to the rear on Foushee St. I uncoupled the hose as John Monk pulled the dead load of 1 1/2" hose and handed me the wye. I connected the 2 ½" hose to the wye which reduces the diameter of the line and provides water to two inch and a half lines. Henry signaled to Glenn Grooms to charge the line and we forced the rear door. Directly inside the door was a staircase. I had the nozzle and halfway up the smoky stairs, fire was burning through the walls. We knocked down the flames and continued to the second floor. On the second floor we couldn't see five inches in front of our faces and kept banging into tables and chairs. The heat was tremendous as we worked our way towards the front of the building. About 20' from the front we ran out of hoseline and the ceramic floor tiles were coming loose from their cement and sliding under us with each mo-

Jake Rixner

tion. I asked John "What should we do?" And he replied, "I think the fire is under us." We retreated to the stairs banging and bumping into the same tables and chairs we had hit on the way in. Downstairs there were several companies working at pulling down the ceiling to expose the fire. The smoke was banked down to about 3' from the floor and below it you could see the boots of the truckies as they worked their pike poles. They were having great difficulty opening the ceiling as it was plaster with a wire metal mesh behind it. It took about 3 minutes to make a small hole and once you got your hook in the hole, it was almost impossible to get it back out. And to make it even worse, you couldn't see the hole you worked so hard to make. Everything was done by feel. We set our line down and doubled up on each pike pole to help the truckies pull ceiling. Within minutes our arms tired and we felt like one of those stretch dolls you buy in a children's toy store. We worked past the point of exhaustion.

The battle opening the ceiling lasted through three air bottles and finally 48 minutes into the job, the fire went through the roof and Captain Emerson withdrew all firemen from the building. It was the first time I'd ever been glad to come outside a burning building.

The second alarm was struck at 0046 hrs bringing Engines 1-12-13 to help us. The smoke at the street level made it necessary to wear a SCBA and even Glenn our pump operator had one on. At 0110 hrs, Captain Emerson struck the 3rd alarm, bringing 15-17-42 into the battle. Ladder pipes and stang guns sprayed the building, but nothing seemed to work right at this fire.

We returned to quarters shortly after the sun came up and we were all exhausted. The "A" shift guys were going home to get some sleep before they went to their Fourth of July picnics and other events, but yours truly had 24 more to work. To add insult to injury, every piece of apparatus that was on the first alarm had black soot baked into the paint. Both our pumpers looked black in color. After line-up we washed the equipment, but the soot remained in the paint. We washed them a second time but the stuff was still there. The Battalion Chief came by on his normal rounds and ordered us to buff the trucks with compound until the red shined again. We finished the polishing about 1500 hrs. and I had the idea that a short nap was in order. The dispatcher had a different idea. We ran several calls and around 1900 hrs we sat down to a nice meal. I helped with dishes and climbed into my bunk and prayed we would stay in for the night. All was well until 0230 hrs when the box opened for a building fire at 1226 W. Franklin St. 5-12-18-1-43-47-53. Something was wrong because 10 and 6 were missing from the assignment. We responded and heard 43 mark on scene with a 3 story building fire on the first floor. When we got to the building 43 was bringing people down ladders from the upper floors. An arsonist had thrown a Molotov cocktail through the back window and the heat and smoke now filled the stairwell trapping the people on the upper floors. Grayson had the pipe and I was backing him up and we pushed the line hard into the three rooms on fire.

We knew there were a lot of people depending on us and this fire had to go out quickly, or some of them might die. We found out later that 10 and 6 were operating in the 2300 and 2600 blocks of Franklin on numerous dumpster, trash, and a motorcycle on fire. Probably set by the same sick man. Grayson was always fun to work with and he could find humor in any situation. After knocking the rooms, the radio began to beep and another building fire was dispatched in the 1100 block of Franklin. Grayson flashed his big grin and said, "That sick fucker is having multiple orgasms tonight." Even though I was dead tired and in my 43rd hour on duty, I had to chuckle at the though of a sick, twisted fucker creeping through the dark alleyways getting his rocks off setting fires. We overhauled the burned out rooms and picked up our dirty wet hose. The hose lines at Five Engine seemed to stay wet and dirty just like the men who worked there.

The next morning I went home and slept for 12 hours, I woke up and had dinner with my wife at 9 p.m. and was due back at the firehouse tomorrow at 0700hrs. Yes, life was very, very good.

Jake Rixner

16 Engine Company of the Pittsburgh Bureau of Fire. Working at a house fire in 1978. As a child I lived 4 blocks from their firehouse. This is the same Seagrave pumper the author remembers seeing in the 1960's during visits with his father.

My first car, a 1965 Plymouth Belvedere. I bought it for $600.00 and waxed it about twice a week. 383 c.i., 3 speed Hurst shifter, Air shocks and glass packs for the motor heads. 8 M.P.G. Passed everything but a gasoline station.

Capitol City Fireman

The "Out of Towners." So-called crew that got the wagon out for over two years. Two men went on to Fairfax County, Va. and the driver became a Fire Chief in California. This is the 1967 LaFrance That I could hear leaving the firehouse from miles away while delivering the Pittsburgh Press newspaper.

Richmond Fire Bureau recruit class of 1982.

99

Jake Rixner

At the house of Refugee Truck Companies. Wagon 1979 Pierce, Engine 1976 Mack.

My love for chop tops continues. This 1982 LaFrance was the last open cab made in the Elmira factory. 1984 at the Ocean City, Maryland parade.

Capitol City Fireman

Just starting out and wanting to be the best fireman in the whole world. Note the rookie glove marks on the face.

Charles Store. Light smoke condition, looked like a trash fire on arrival. Less than 5 minutes later we were struggling to keep these fire trucks from burning.

Spring of 1983. Engine Co # 5 "B" Platoon. Using a reserve engine. We were running as a triple combination as both our rigs were in the shop. L to R Me, Lt. Pulliam, Tupp, Duke, Herman. Gene-O is taking the pic.

The reason I worked so hard my entire adult life. My family: Julie, Katie, Jimmy, and me in front of #4's Hose Wagon. Mack wagons were like battle tanks, they took a beating and never seemed to breakdown.

Captain Richard Emerson of #5 Engine talks to a reporter after a fatal house fire on Rose Ave. Our Company was caught in a flashover that morning. The action didn't get any better than #5 Engine Company in the 1980's.

Jake Rixner

8-alarm fire at 14th and Cary Streets resulted in the shortest part time job I ever had: 3 hrs. I left for lunch and saw the smoke and never went back. The call back crews had to search for equipment, and never did find air packs.

In my 8 years at #5, I went to at least six working fires in this building alone. We averaged about two to three runs there a week. The rooms there went for just a few dollars a night.

One of the last good jobs at the Earle Hotel. The Five Diamond Jefferson Hotel was in the next block and applied pressure to have the building razed.

Main Street Station fire October 1983. We burned the roof off it. Fortunately, the building has been restored to it original state.

Jake Rixner

Charles Store. This fire taught me to respect vacant buildings, they can become a death trap very quickly.

Charles Store the next day. Fire is a force of nature that needs to be controlled. Some of the best firemen in the U.S.A. couldn't stop this destruction.

Capitol City Fireman

Storer Hall, Virginia Union University. Smoke damage to one of the dorm rooms. 100 students were rescued with ladders by Number 3 & 6 Truck Companies. The only time in my 32 years of firefighting I have seen every ladder stripped off of two hook & Ladder trucks.

Capri Restaurant. The only time I worked 48 hrs. straight and we spent the entire 1st night here. Also the first time in my life I was happy to be ordered out of a fire building.

Jake Rixner

Wires burning off a row house. I don't care how many times this happened to me, it always scared the hell out of me. To this day I hate electricity.

Paper warehouse 14th & Mayo. Yes, that is a 4" line going inside. We used a stang gun and 2 two and a half's to keep fire from an entire wing of this warehouse. That's the beginning of Lake Griggs at the loading dock.

Capitol City Fireman

Repeat after me; "The roof, the roof, the roof is on fire!" Another job on Broad Street.

We became experts at fighting fire in two story wood frame houses, or as Grayson used to say, "Two story flames."

109

Jake Rixner

Mike Edmonds's first fire on the pipe, Goshen Street. I'm on the ground helping Earl from 10 pick up their line after the fire. Fire was in the rear bedroom 2nd floor. Notice the structural condition of the bricks in the arch above the front door.

Capitol City Fireman

Gary and me at West Lanham Hills, Maryland Summer of 1979. Gary was the best man at my wedding. He helped me at the attic fire in Wall, and has always been a good friend. He went on to make a living fighting fires for the citizens of Va. Beach, Va.

My father, my sister Loretta and me circa 1968 hanging onto my father's coat tails.

111

Engine Company 281 West Lanham Hills, Md. Me, Sully, Curtis, and Richard.

The three Lieutenants; Me, Ed Smith and Pete Lund.
*Photo courtesy of Curtis Phillips

- Chapter 12 -

Acting Lieutenant

Our cowardly leader came back from his leave period but apparently didn't grow any balls during his time off, and we had a couple of more "scenes" during that time including one of the only times I voluntarily gave up the pipe during a fire. We went on a second alarm to Barton Avenue on a particularly hot day in July. I was driving the wagon and we laid a line to side 3. Arriving in the rear of the fire building we could see that the fire had been out every window on both floors, but was now knocked down because the first alarm companies had done one hell of a job and needed some fresh troops to come in and help open up walls and ceilings. We started to break the line down to the smaller inch and a half when he demanded we put the two and a half inch nozzle on it and take a position in the rear yard. Once we got water he ordered we spray the water into a burned out window on the second floor. We had a fill-in man working that day and I handed him the pipe and took the last position on the line. I was so fucking mad I couldn't see straight. It is a cardinal sin to throw water into a building when men are inside. I had to laugh when other officers came out of the building screaming at this fool. Luckily he was high up on the captain's list and the city soon promoted him. He wasn't much of a lieutenant, so promote him to captain. I know, it sounds crazy, but it was a great deal for us, and this was the way the City worked in the 1980's.

With our boss gone and no one appointed to take his place, we took turns "acting" as the officer for the company. Acting was a good chance to learn the leadership skills needed to be a good fire officer. I remember my first time in charge; it was August 18, 1983. That day was busy as ever as we ended up with 11 runs. 4 of which involved smoke in buildings. Saturday 8-20-83 was more normal with 5 runs, an automobile fire, two brush fires and two dumpsters. My third day acting convinced everyone that I was a black cloud (someone that fires and weird incidents seem to follow.) Monday 8-22-83 started off quiet, with only one run until 1524 hrs. A strong band of thunderstorms swept into the area and the radio was beeping and chattering constantly. We were returning from MCV when the assignment for 4300 Dover Rd. came in. Herman was driving the wagon and when I asked him if he knew where Dover was, he just smiled and said "Delaware." We pulled over to the curb and I went back and asked the others if anyone knew where it was? Blank stares told me they didn't know. We drove to the nearest 7-11 store with out lights or sirens and I went inside and bought an ADC map book. We still arrived at the large church on Dover Rd. before anyone else, and luckily it was an alarm malfunction caused by the storm.

Clearing from there we were sent to Lombardy & Leigh Streets for a building collapse. A shopping cart manufacturer was located at this corner, and a strong gust of wind had toppled a two-story brick wall onto the adjoining one story flat rooftop causing it to collapse. Upon our arrival we encountered injured workers, buried in the collapse, arcing electrical lines and a host of other hazards. We entered the building and began removing the injured. I had the foreman with me and asked that he secure the electrical service in the building. He smiled and explained that because of the vats of acid and caustics, even if he killed the power now, because of the cycle of the conveyer it would take 30 minutes to shut off the electricity. We treated the injured and sent them to the hospital, then began the long wait for the building inspector. The rest of the tour was filled with wires arcing calls and false fire alarms. The day sheet for that tour lists 16 runs with one building collapse and one small fire in a church.

Acting as the boss for 24 hours was fun and rewarding, but it was also like babysitting for your younger siblings, yes you are in charge, but not really. It was also interesting how the responsibility changed some people. Some of the laziest firemen became superman when task with the acting position. Power truly intoxicates some folks.

September of that year Julie found out that she was pregnant with our first child. I was happy that we were starting our family, but the idea of becoming a father at 21 was also scary. Little did that kid know, he would be going to big fires even before he was born. We were sitting down to eat dinner in our sparsely furnished apartment when the scanner announced a second alarm for 1504 Belleville St. The temperature was in the 90's and I could tell by listening to the radio that the fire was large. When they put the 3rd alarm in I asked Julie if she wanted to come with me. We drove down to Scott's Addition and parked about three blocks away. We were able to walk all the way up and stand directly across the street from the one story paper warehouse. Fire was through about 25% of the roof and things didn't look good. The "A" shift at 5 Engine had two and a half through the front door and you could see the sheer exhaustion when the guys came out to change air bottles. Lt. Pulliam was working and when he saw us standing across the street, he asked me if I wanted to work. "Sure," I told him, and then remembering how I had driven the engine the day before, I thought maybe my gear was still in the compartment. I ran up to the corner where the Mack was hooked up to the hydrant and sure enough my turnout gear was exactly where I'd left it. I grabbed an SCBA and joined "The "A" Team inside the huge warehouse. Henry had the pipe as Grayson, Clarence, and I struggled to feed more of the heavy two and a half inch hose forward. Everywhere we looked from the 1 1/4:" tip in front of us was an orange sea of flames. The 326 gallons per minute were doing very little in beating back the fire.

Little by little we were making slow progress, but it seemed like every time we came out to change air bottles, we had to fight for the same ground when we returned to the inside of the building. On about the fourth trip outside Henry collapsed from the heat. Grayson and Clarence were taking oxygen from 3 Truck and then Lt. Pulliam collapsed. I helped the ambulance crew load Lt. Pulliam into the ambulance. He didn't look good. His skin color was ashen, and he looked like he was having a heart attack. When they closed the ambulance door, I didn't think I'd ever see him alive again. Before it was over the entire crew was taken to the hospital except for John Monk, and me. The fire ended up with 5 alarms and 11 of our 20 engine companies saw action and 4 of our 9 trucks worked the fire. After being relieved I went across the street to pick up Julie. Captain Butler was acting 53 and standing next to her just watching the fire. I was a little shook up and Butler asked me, "What's wrong?" "I think Lt. Pulliam just had a heart attack" I replied. "Good, serves that fat fucker right." Butler commented and walked away. Captain Butler and Lt. Pulliam were close friends, and I knew Butler loved to play, but I couldn't believe my ears. I just shook my head and walked away.

Jake Rixner

There are a million Captain Butler stories to tell. Kenneth Butler was an old school, no nonsense fire officer with a quick whit and a sharp tongue. He was in his mid 50's and had been a city fireman since 1956. He could cut you to ribbons and be in the next room before you knew you'd been had.

On his days off he owned a small construction company and was a bricklayer by trade. The kitchen door at station # 3 faced the alley. Directly across two vacant blocks was the Fire Chief's office in plain sight. The fire chief was Ronald Lewis; Chief Lewis had retired from the Philadelphia Fire Dept. and come to work in Richmond. He was the first black fire chief in the history of Richmond. Both he and Butler were strong willed men, and sparks soon flew.

One of the firemen from 41 brought an African blow dart gun to work; we had put up a target on the bulletin board next to the back door and were taking turns practicing with the gun. Now it is against regulations to have any kind of weapon in a city firehouse, but Butler didn't care as long as we didn't put each other's eyes out. The back door opened just as a dart hit its mark as Chief Lewis entered the building. Butler was sitting at the kitchen table reading the afternoon paper and stood up and took the blow gun and offered it to Chief Lewis and said "Here Chief, you want to try this, you ought to be pretty good at this." You could see the anger in Lewis's eyes and he spun on one heel and went back out the door.

Another time Butler was sitting on the bench in front of the firehouse when Lewis pulled up with the newly elected Mayor. Lewis gets out of the chief's car and Butler continues to sit on the bench. Lewis introduces Butler to the mayor, "Mayor West, I want you to meet the worst Captain in the Fire Bureau, Captain Butler." The two men shake hands and Butler says, "How are you doing?" like it's no big deal to be insulted by the fire chief.

Another time we were getting on an elevator to investigate smoke on an upper floor of a high rise building when just as the doors start to close, an elderly lady comes rushing up and grabs the doors, enters the car with 5 fully dressed firemen. Butler was the ranking officer and we were waiting for him to throw the lady off the elevator when she looks at all these dirty firemen and says, "Geesh, all these men, am I going to be alright?" Butler eyes her up and down as if checking her out and said, "I don't think you don't have anything to worry about." We could barely keep from busting up laughing.

Butler had a guy on his shift we will call Paul. Paul couldn't get to work on time to save his ass. Butler would write charges on him at least once a month but Chief Lewis would slap Paul on the wrist and send him back to the firehouse. I was sweeping the apparatus floor one day when Paul was late for about the 50th time that year when Butler called me into his office. "Read this," he said. I read the departmental charges against Paul, which outline his tardiness. In the comment section Butler had typed, "I know you will not do anything with him, but I'm doing my job anyways." "Jesus Captain, you and Chief Lewis must love each other." I said, and we laughed together.

Gene-O loved to play practical jokes and one afternoon we loaded Captain Butler's locker into the service truck and sent it to 24. The next morning at the kitchen table Butler walks in wearing civilian clothes. He starts looking through the newspaper when he says to no one in particular, "$422.18." We were looking at each other like, what's he talking about, when he repeats, "Four hundred twenty two dollars and eighteen cents." "What are you talking about Captain, " Eddie P asks. Butler still looking at the paper says," there was $422.18 on the top shelf of my locker when I last saw it, it better still be there when it comes back." You could have bought Gene and me for a nickel, as $400.00 was a ton of money back in 1983. In fact it was more than we made every two weeks. And the Captain never locked his locker. I could picture some slug at 24 taking the $400.00 just to teach us a lesson. Gene got permission from the Lt. to use the service truck and set a land speed record crossing town to get to 24 and get the locker back.

Another time when Butler was on the car as acting Battalion Chief, we took his keys to the red car and put them in a pot of water, and placed it in the freezer. All was well until we picked up a run for fire in a paper plant with 53 due as the chief. We were about half way to the run when I remembered 53's keys. "Damn!" I said. The other guys on the back step just looked at me and I explained to them what I'd done with Butler's keys. "Your ass is done," Gene-O said, and when we got to the run Butler pulled up a couple of minutes later driving the service truck. When we got back I surreptitiously thawed out the keys and placed them on the kitchen counter. Butler never said a word to anyone about it. He could take it as well as he dished it out.

Chief Lewis finally exacted his revenge and transferred Butler to 9 Truck. 9 Truck was one of the slowest companies in the city, running about 120 runs a year. Shortly after he went out there I ran into him at a lumberyard off duty. "How do you like 49? I asked him. He just groaned and said it was so slow it took two hours to watch the TV Show 60 minutes. We laughed together, but I could tell the transfer had hurt him. Chief Lewis had taken one of the best tactical command officers, and put him out to pasture.

In October one of the firemen got a bus trip up to Charlotte for the NASCAR race. Captain Butler talked me into going on the trip. It was all city firemen on the trip and was more like a stag party. We worked the night before and had a fire in a row house on Catherine St. about 0100 hrs. Pinky and I ended up on the line and knocked two rooms on the second floor rear, the normal overhaul and final bath finished the job and put us back in quarters about 0230 hrs.

About 0300hrs we went back to the same building to find the entire 2nd floor on fire. The arsonist had returned and really had it going this time. We met for the bus at 0600 hrs and I wanted to sleep on the way. Captain Butler wouldn't let me rest. Every time I closed my eyes he made some request. I asked him to let me get a nap. He looked shocked and said, "You told me you were tough enough to hang, so let's go big boy." We played dominoes and drank heavily the entire way to Charlotte.

A couple of years later, we were up playing dominoes in the kitchen at 0300 hrs when we heard companies start going on air. There was a flood in Richmond and some genius had gotten the idea that the fire companies could mark the high water lines when the flood crested so studies could show how deep the water had gotten. The game was over and we were on our way to the bedroom when I heard Butler on the radio. "Hey guys, hold up a minute, I just know Butler's going to say something priceless." Sure enough, "49 to Radio." "Come in 49." The dispatcher answered. "You ain't going to believe this, but we are going out to paint lines on the ground!" A couple of minutes passed when he came back on and said. "Near the water there are gravels, the gravels is wet and might be slippery. Do you want us to take a chance and step on the gravels, and possibly fall in the water?" The dispatcher is laughing and can barely answer, "No 49, we don't want you to take any chances." "O.K. we are 10-8 returning to quarters then!"

Captain Butler retired in 1988 and continued to lay brick well into his 70's. He came down with a rare cancer of the brain and died in November of 2001. A couple of weeks before he passed away he was sitting on his porch with his son Marc, there were two buzzards circling overhead and Butler in his typical straightforward manner nodded to the birds and said, "They are a little early, aren't they?" They don't make fire officers like him anymore, God I miss him!

Grayson Finner was another one-of-a-kind coworker. He was a light skinned black guy in his 30's and during cold weather wore a gray leather helmet liner on his head that looked like something like a World War I bomber hat; the kind that Snoopy wore on Peanuts. He always had a huge grin and could be counted on to break the ice at just the right moment in the heat of a battle.

Grayson had an irreverent sense of humor and especially hated officers breathing down his neck. Anytime he thought he was being supervised too closely he would launch into his "YESSEM BOSS" routine. This drove the black supervisors crazy, and scared off most white ones. He was about 5' 10" tall and was as agile as a cat. I remember being sent down to relieve companies operating at a large fire on Jefferson Davis Highway at 2300 hours one cold December night. This company made the foam rubber that is used for automobile seats, and sofa cushions and the stuff that came out was about ten foot high and twelve feet wide. It was like walking on a giant sofa cushion. The fire was contained but required our presence through the night. The original companies had been here since 3 pm and we just took their engine out from the hydrant and hooked ours up using the halfway frozen hoses already in place. We manned the hose lines in a warehouse and got soaking wet. Temperatures dropped into the single digits and the diesel engines couldn't produce enough heat to warm the cab during one break at 0300 hours as we huddled together, 5 of us in a 3-man cab. Grayson produced a 12 pack of beer and began passing them out. I opened the can and took a sip, but it was too damn cold to drink a beer. I asked him, "Where the hell did you get beer at this hour?" as it was about 3 a.m. and stores quit selling beer at midnight. He just grinned and said, "Don't worry Rix, I've got my ways."

Another night he came to me when I was acting and said, "Rix, I gotta family emergency, I'll be back in a while," and rushed off. We ran with 5 or 6 men depending on manpower so it wasn't a big deal, but I wondered what his "emergency" was, and knowing Grayson, it could be anything. He came back two hours later and reported that his roommate had been working on his motorcycle in the living room of his apartment. The bike caught fire and burned out 6 units in the building.

Grayson lived out in the county, and this county isn't known for being aggressive on the fire ground. He reported to us with his typical smile that his building had a big hole in the roof. "It looks like Pac-Man took a big bite out of the roof."

But the night that I came to see Grayson for the kick ass fireman that he is, was on January 2, 1984. It was a normal tour of duty, the average citizen might not understand how any day in a firehouse is normal, but after a while you get used to the normal, automatic fire alarm runs, food on the stove, auto fires, garbage fires, mattresses, etc. It is a pace or a rhythm that you pick up on after a while. None of these are exciting, all of them are dirty and they just become a part of the normal day for any city fireman.

At 1901hrs the box opened for 6-5-10-33-41-53 to 12 East Main St. for a building fire. This address is only 3 blocks from 6 and soon they were on the scene with smoke showing from a three-story brick and joist apartment building. I was driving the wagon with Gene-O in charge. Grayson was on the back step with a fill-in guy and Herman driving the engine. We arrived 2nd and were told to stretch our line to the third floor. The fire was in the walls on the 2nd and 3rd floors. When we got up to the 3rd floor visibility was nil and because there was only one ladder company assigned to this fire, we had to open our own walls to expose the fire.

We were working for about 10 minutes when Battalion Chief Tyler tapped me on the shoulder and asked, "How's it going up here kid?" I noticed that he had a turnout coat on but was wearing his normal dress uniform cap, and no air pack. I figured that if Chief Tyler didn't have a mask on it must not be too bad up here, and removed my face piece. I immediately began to choke and I believe I would have died right there if it wasn't for a near-by window I stuck my head out of to allow me to get my mask back on. These old school officers always impressed me with the way they were able to breathe that smoke.

After the fire we picked up our hose and returned to quarters for about 45 minutes when the box opened at 2213 hrs and sent us to 409 North Boulevard for an attic fire. The assignment was 18-12-10-5-47-43-52. We were running 4th due engine and I didn't expect much action, but when we arrived Chief Hughes ordered us to take a hoseline from 12's pumper into the attic. I had the pipe with Grayson backing me up and we climbed through a scuttle hatch, the smoke was thick and the heat moderate as we crawled from rafter to rafter towards the back of the building when something shocked the hell out of me. Grayson was patting me on my leg and asking why I stopped but I kept getting shocked every time I moved. Grayson was getting pissed so I just gave him the pipe and when he shoved past me, he got lit-up too. "God damn Rix, why didn't you tell me you were getting electrocuted?" "I figured you'd find out for yourself," I replied. When the smoke cleared we could see the place had post and terminal wiring and bare wires that ran from porcelain terminal to terminal.

They teach you not to squirt smoke, but from the middle of that attic we could hear but not see the fire. Being shocked stopped our forward progress so we opened the pipe and let the water flow. It worked and the fire went out. Chief Hughes told us to return to service with-out picking up the hose we used. Damn this was like my Uncle Schmo in

Pittsburgh. He was on a squad and that's exactly what they did, arrive in a van, use someone else's stuff, put the fire out and leave. Leaving the fire we were enroute home when the dispatcher called and ordered us to transfer to 10's quarters. We backed into 10 and were in the process of making a pot of coffee when the box opened up at 2258 hrs. and sent us to an apartment fire a couple of blocks behind our own station. 8 West Charity St. had fire on the 2nd floor with people trapped on the 3rd. We slid the poles and hit the street fast. We had people trapped on our third working fire of the night and I had the wagon stretched out as fast as it would go. We turned into Charity Street practically on two wheels and could see heavy fire shooting out of the 2nd floor windows. Gene-O marked on scene, and as we pulled up a woman jumped from the 3rd floor and landed 5 feet from my door with a snap and a thud. The snap was her leg bones breaking, the thud was the rest of her hitting the ground. I looked up to see another guy following her out the window. I yelled for him to hold up and Grayson climbed up on the porch roof with his bare hands and Gene passed him a ladder he used to climb from the porch roof to the 3rd floor.

Remember, fire is blowing out the 2nd floor windows like a torch and Grayson has his ladder to the left of the windows, and leaning and reaching to his right he makes contact with the guy and guides him onto the ladder. While he's doing that I grabbed the trash line and tried to knock the flames back into the apartment through the window directly below them. Once Grayson and the man are on the ground we hand stretch a line back to the hydrant and another to the second floor. Grayson had the pipe and we easily pushed through the entire apartment running on pure adrenalin. When we came out of the building, the street was full of news reporters, buffs, volunteer firemen and others who had heard the dispatchers tone of voice and heard Gene-O's on scene message. For the public housing projects of Gilpin Court, it was incredible.

The next workday I was driving the engine and I hated driving the engine as it was the piece we hooked up to the hydrant and supplied the water to the guys who got to do the fun stuff. The engine was also normally a block or so away from the fire. Sometimes you didn't get to see any action at all.

At 16 minutes after midnight on January 5, 1984 the box opened for 2610 Monument Ave. for a building fire. We came down Mulberry St. and the wagon laid out from the same hydrant as 18 Engine. 18 was a single piece engine and I had Doug Bracey from 18 helping me put my engine in to the hydrant. As soon as we got hooked up 18 and 5's wagon called for water. I charged both lines and was monitoring my gauges.

From my position in the middle of the block on Mulberry I couldn't see anything but smoke rising against the streetlights in the next block. At 0031 Hours Chief 52 put in a second alarm. Engine 20 laid another line from my pumper. By 0100 hrs things on my end of the job had slowed to a boring crawl. I got into the drivers seat to try to keep warm but was bored out of my mind. A little after two a.m. when a popular nightclub called Much More closed, the drunken patrons began to make their way home. I was sitting there trying to stay awake when a pretty young girl came staggering down the block. I got out and was trying to look busy at the pump panel when she said, "Hey fireman, tell me some exciting stories." I told her I didn't have any exciting stories as I'd been standing near this hydrant all night. Without missing a beat she said, "I got a dog that does the same thing." Like I said, driving the engine wasn't my thing.

The run for 2832 Monument Ave. came in at about 2 in the morning and I was driving the engine that night. Lieutenant Bridges had been transferred in from 6 and he was one of the best officers I have ever worked with. He was an older white guy, and had come up the ranks in Church Hill, was a good fireman with a good background and a good reputation. He started off at 2 Engine, which doesn't even exist any more. It was closed for budget reasons during the 1970's. It's kind of strange, you run hundreds and hundreds of fires and some buildings you never go to at all 2832 was one of these buildings until May of 1985 when we went there 2 times in a month. Both times we had working fires on the top floor. It was a turn-of-the-century 3 story walk-up; huge apartments with 12' plaster ceilings. The fire escapes on sides 2 and 4 still had the wooden ice boxes, where back in the days before refrigerators; the iceman would come up the fire escape and set the huge block of ice into the building from the exterior into the "ice box."

It was a historic old building and it was a shame to see, but it was becoming obvious that someone was trying to burn this building down. What I remember about 2832 was that I was driving the engine both times and was located about a block away at Grace Street pumping the water to the fire. The fire marshals later arrested the buildings security guard and the fires stopped.

- Chapter 13 -

Fun with the Cops

Police officers and firefighters in Richmond got along very well for the most part. Policemen usually saw the firehouse as a safe haven; somewhere they could stop on a hot day and enjoy some air conditioning, use the bathroom, the telephone, and get a cold drink. However, the people who work in firehouses are usually pretty tight-knit groups and policemen are, as a rule, pretty strong individuals. It made for some interesting times. That's why I call this chapter "Fun with the Cops." We had many different policemen come into our firehouse, and all of them had to be educated quickly that the firehouse is a place where the majority rules, not the individual.

I want to mention a few of these visitors. One of the first police officers that I met was a guy named Al, who we affectionately called "Porky" because had an upturned nose and a round face and looked similar to Porky Pig. Al even called himself Porky! Al was constantly in trouble with his supervisors and when you are in trouble in the police department, you get assigned to the paddy wagon. Driving the paddy wagon was not looked upon as a good duty to have for several reasons. The paddy wagon picks up all the prisoners, transports them to the lockup. This duty usually involves a long day of transporting a whole lot of drunks who invariably throw-up or do even worse in the back of your wagon, which has to be cleaned up by the officer. More than once Porky would pull in behind the firehouse, grab the garden hose, and proceed to decontaminate his wagon. One of the funniest stories I re-

Jake Rixner

member about Al happened late one night when we were at a trash can fire behind a high school which had a wooded area behind it. We found Al asleep in his paddy wagon (a policeman's shift after about 3 or 4:00 in the morning starts to slow down) right near the woods. Normally there would be two or three police cars with him, and one guy would stay away while the others got a little cat nap.

On this particular night, Porky's friends had left him and he was asleep next to the wood line, alone in his paddy wagon. After we extinguished the trash can fire, we searched around to find a way that we could mess with our buddy Porky.

An abandoned tire was soon discovered, and at the time the paddy wagon only had one small bubblegum warning light in the middle of the roof. We quickly set up a step ladder and proceeded to put this abandoned tire flat on the roof of the paddy wagon, surrounding the warning light. It was all in good fun. We quickly and surreptitiously rolled out, leaving our buddy Porky to suffer whatever fate awaited him. In a plan that couldn't have been scripted any better, Porky arrived back at the police precinct at 0700 hours for shift change where there were numerous police officers and supervisors in the parking lot who observed Porky pulling in driving the paddy wagon with a junk tire perched upon the roof, blocking his warning light. Porky's sergeant walked up to him (probably with a smile on his face) and proceeded to ask Al if he had been asleep. Porky replied, "Oh no Sergeant, I haven't' been asleep, I been out on patrol."

"You sure you ain't been asleep, Al?" the sergeant asked again.

"Oh, no sir, no sir, I wouldn't think of sleeping on duty," Al responded.

"Well, step out of the wagon, then and explain to me what that tire is doing on your roof, Al.," the sergeant said.

Well, as you can imagine, you could have bought Porky for a nickel! To the delight of about 20 fellow police officers who were observing this fiasco, Porky was in some hot water.

The next day, Porky came into the fire house accusing us of committing such a dastardly deed. Of course, we all denied everything. For years, Al was left to scratch his head and wonder who could have possibly put an old junk tire on the roof of an 8' tall Dodge van, as he slept inside.

Before I talk about our next police friend, I want to make the point that most police officers start off like any other working man, just trying to pay the bills and trying to do what is right, but the stress and the aggravations of the job lead to some interesting things. As I said the firehouse is usually run by majority rules, and police officers being individualists often run into a little trouble. There was one policeman who would come into the firehouse all the time to use the bathroom; he wasn't the most personable type of guy, and he was easily upset, so naturally he was the perfect mark for some firehouse fun. This guy happened to be a friend of one of the firemen on the A shift, and it was his custom to come in and take off his gun belt when he had to use the bathroom. He would proceed to go into the bathroom and sit on the toilet with the stall door closed and his gun was out of his sight. Naturally, firemen being the way we are, it wasn't long before we were taking his pistol out of his holster. One of the fellows, I think it was Gene-O, took all the bullets out of the gun and proceeded to tuck the empty pistol into his waistband. In the interest of safety, he showed a couple of us that all 6 cylinders were indeed empty and that there were no bullets in the gun. Well, when our police officer friend, I'll call him Joe, finished his business and came out of the bathroom, he quickly noticed that his holster was empty.

There is nothing worse that a policeman can do than lose his firearm, especially in a building full of 10-12 crazy guys. It was about the time he noticed his holster was empty that a loud scream came from the apparatus floor. Joe hurried to the apparatus floor and observed one fireman running full speed being chased by another fireman who had Joe's pistol in his hand, yelling something about, "you mess with my wife, you son of a bitch, now you die!" Just as Joe yelled, "Stop!" the man began to shoot the (empty) pistol at the other fireman, only to hear the cylinder to go, "click, click, click." Just as Joe was on the verge of having a heart attack, all 10 guys in the firehouse erupted in laughter at Joe's expense. Needless to say, that was the last time we saw that uptight son-of-a-gun. He quickly gathered his equipment and left Engine 5 never to be seen there again.

Back in those days, the police chief had been on the job for over 40 years and although for all intents and purposes, he was a good police chief, he had some strange, old-guy type quirks. One of these was a administering a severe punishment for any police officer who got out of the patrol car without putting his dress uniform cap on. For some reason, according to this chief, not having a dress cap on when getting out of the patrol car was considered to be "neglect of duty." When coming into the firehouse, one of the first things the police officers did was take their hats off and toss it on the table.

By the time they went to the bathroom or as soon as they got out of sight, the hat was quickly placed in the freezer. If the policeman was not well-liked, the hat was first rinsed off in the sink before being put into the freezer, which would give him a frosted head. It got so comical that several policemen would come into the firehouse and go directly to the freezer and toss in their hats to save us the trouble! It was all good fun. We had a lot of fun with the police.

Some policemen were rampant individualists who are stuck on themselves. There was one guy; I'll call Blackman, who was a canine officer. He used to come into the firehouse and usually his trained dog was placed upon whoever was on the watch desk and was told by Blackman to watch him. So if you happened to be the individual sitting watch when Blackman came in, you were essentially stuck at your desk while Blackman went back to the kitchen to have some coffee and shoot the breeze with the rest of the members.

This dog was so well trained that you literally couldn't get up to use the bathroom, you were stuck in your chair because as soon as you moved the dog would start to growl and he was so menacing you just knew that he would bite you if you moved. Blackman was very proud of his German Shepherd and I don't remember the dog's name, but I do remember how well-trained he was. We had a guy named Fenner on our shift who would try to pet the dog and of course, the dog would growl and bark. Blackman was quite proud of this and would boast that, "Nobody can pet my dog, he's that mean." Fenner kept telling Blackman, "I can pet that dog." Blackman would continue to come by the firehouse about twice a week.

During one of his visits, Blackman went to the apparatus floor and saw Fenner and his dog. Fenner was holding the dog's penis and was masturbating the German Shepherd. Blackman was so outraged that he grabbed the dog's lead and tried to pull the dog away from Fenner, at which point the dog looked up at Blackman and growled to the delight of all the firemen in the house. Well, when Fenner had finished pleasuring the dog, Blackman left the firehouse cussing and fussing, once again like his counterpart, Joe, never to be seen at the firehouse again. He had been embarrassed at the hands of the firemen on duty.

I could talk about many, many more cops and the fun we had, but one guy especially stands out. Charles Bisher. Charles, a nerdy looking white guy complete with everything except the pocket protector and slide rule, came into the firehouse one day and announced that his name was Charles Bisher. We responded to his statement with the greeting, "Hey Chuck!" Charles became outraged, his name was NOT Chuck, it was Charles and he quickly made it known that he did not want to be called Chuck. Every good fireman knows that this meant that Chuck instantly became his name. In fact every time Chuck came into the firehouse, it was announced on the public address system throughout the firehouse, "Chuck's here, Chuck's here!" Most times this would enrage Charles and in his anger he would make comments that would amuse everyone involved.

The Chuck nickname held until one hot summer night we were called to take a report of attempted arson. I was driving the wagon that night, Valjean was lieutenant and we had the regular complement of guys for a two piece engine company. We drove over to Gilpin Court to take the attempted arson report, responding with no lights, no siren at 3:00 am. When we arrived on Hill Street, there were about 300 people in a courtyard, hootin' and hollerin', drinking and carrying on. As we pulled up, Valjean looked at me and said, "Y'all just stay on the wagon, I'll get the information and we will get the heck out of here." As Valjean got off the equipment, Gene-O got into the cab and we were sitting there shooting the breeze. We were parked directly behind one police car and about that time another, unmarked police car pulled up on the opposite side of the street. Remember there were about 300 people in this courtyard. The woman identified a man in a yellow shirt and stated that he tried to set her apartment door on fire. Officer Charles Bisher enters into the sea of people by himself, and proceeds to arrest the accused party. As he is bringing his prisoner back to his patrol car to wait for Porky and his paddy wagon, gunfire erupts about a block behind us. I then slide down low in my seat, glance in the rear view mirror of the wagon and I see one black man running up St. Peter Street as fast as he can turning around and armed with a pistol shooting behind him as he is running away from someone.

A few seconds later, another black man, also armed with a pistol appears and is running facing forward, shooting as he is running. It looked like a scene of playing cowboys and Indians, only this was with real guns and real bullets. As this evolves, Chuck, I mean Officer Bisher had just put his prisoner in the back seat of his patrol car and was awaiting the arrival of the paddy wagon. Charles then runs up to my driver's door and tells me, "Watch my prisoner!" and he runs up the middle of the street. What I observed in the mirror made me chuckle.

While the sidewalks were relatively dark, due to the foliage on the trees, in contrast, the middle of the street was well-lit by street lights so as Chuck was running down the street he was in plain view, screaming into his radio, while reaching for his pistol. He made several grabs for his pistol and came up empty about a half a dozen times, but when he finally was able to retrieve his firearm, it came up, holster and all from his belt. I was laughing at this comedy of errors and made the remark to Gene-O as both gunmen were disappearing up St. Peter Street, "If they start coming this way, we're leaving!" he said. "What about the Lieutenant?" I replied, "Oh, he'll find his way back."

Remember the unmarked police car? Tucked conveniently inside of the car was a small black female officer who proceeded, as the gunshots erupted, to jump back into the car. At first, we thought she was going to race down St. Peter Street to chase the 2 gunmen, but we quickly realized that when she jumped out and then back into the car a half dozen times with a wide-eyed look that she was scared to death and had no intentions of interjecting herself into this running gun battle. In fact, Gene-O even said, "Look at that bitch, she's scared to death." I told him, I said, "Gene, she's collecting a paycheck every week, isn't she? That's her job; she needs to go down there." After a few minutes had passed, police cars filled the area responding from everywhere. Soon Officer Bisher returned with another criminal; he had captured one of the gunmen.

From that day forward at Engine 5 it was, "Officer Charles Bisher is in quarters," and he was never again to be called Chuck. That is just some of the fun we had with cops.

- Chapter 14 -

D.C. Fire Department Calls

In early 1985, I received a letter in the mail from the D.C. government for a physical examination for the position of fire fighter. I had pretty much forgotten about D.C. in fact I was pretty happy in Richmond, going to plenty of fires. The idea of a free comprehensive physical that included a chest x-ray, blood work, and everything else appealed to me. And since I was only 22 years old, I really didn't go for a good physical every year and this sounded like a good opportunity.

I made the trip to D.C. and stood in line at the Police and Fire Clinic with probably about 200 guys waiting to see the doctor. We were all wearing only our underwear and a pair of paper slippers. One of the guys behind me in line, we stood in line for a while and just made small talk, ended up becoming Battalion Fire Chief Keven Sloan and we have been friends for many years.

About a month later, almost out of the blue, having not had an interview or anything, I got a letter informing me that I had been appointed to the District of Columbia Fire Department and was to report on August 19, 1985, dressed in a suit and tie, to the District Building, to be sworn in as a District of Columbia fireman. I spent a few weeks struggling with whether or not I wanted to take this job, but ultimately the opportunity to be able to retire after 25 years of service is what swayed my decision. The geographic size of D.C. is about the same as the city of Richmond, but the population density is much higher.

Jake Rixner

This really started off one of the best summers of my life. I met some people; a pretty incredible group of men assembled together as a recruit class, in fact, everyone in that class was born to be promoted in one way or another. Each morning we started off by walking in the front door in full uniform, signing in the log book, taking our fire equipment and placing it in neat little rows on the drill yard.

After a formal line-up inspection, we quickly changed into PT clothes and started about one hour of torture each morning. The normal drill was a run in military style formation with Zeke, Willie or Danny Troxell calling out cadence.

After about a couple mile run, we would break out of formation, dress in our turnout gear as quickly as possible including SCBA with air on (breathing air) and we would line up and do wind sprints, bear crawls, and a number of other excruciating exercises.

One of the things I enjoyed about the D.C. fire department was the opportunity to train with this incredible group of men. Of all the training classes I have taken in 30 years of fire service, this was by far the best in every way as far as teaching the new guy what he needed to know regarding pulling hoseline, throwing ladders, and using breathing apparatus.

To this day, this training experience remains far and above as the best basic fire training that I have ever taken. One of my favorite things about the D.C.F.D. was that the recruits would practice on a specific skill all week long (this class was only 10 weeks long) and on Friday the instructors would pull out the stop watch. Now, I have always been a pretty competitive person, but the fact that I was competing against my classmates for time, really, really put the icing on the cake for me.

During the team exercises, like ladder exercises, I was fortunate enough to be paired up with John Slavik, Danny Troxell, or Greg Dipski. All these gentlemen turned out to be great firemen and are very competitive by nature so, needless to say, we didn't lose too many events. In fact, we set some records.

We had practiced all week long climbing the face of the 7-story tower building using a device called a pompier ladder. A pompier ladder just hooks into a window and has one metal spur with spikes protruding from it (like what you see on telephone poles) and you hook the ladder in the window and proceed to climb up the spur using the spikes as rungs. As the ladder sways back and forth it seems like a precarious situation, but actually it is an excellent training device to use to teach a young fireman how to climb ladders. On Friday, I was in the lead for setting the class record on the pompier ladder, until Danny Troxell beat me out by 1/10 of a second. I came in at #2 at that event. Another interesting event aside from challenging ourselves and learning what we were made of one of the more interesting events during our class was after coming back from PT one day, we noticed smoke coming from the smoke house (the 3-story burn building) on the academy grounds. As we dressed and got in our normal line up, there were two 2½" lines stretched to the building from side 1 and 3 (the front and rear) of the building and the instructors quickly split us into two groups and sent one group to the front and the other to the rear with no further instruction. When we got to the door of the burn building, the fire was on the second floor, they bled our air bottles down to about 200 psi of air and slapped us on the back and told us to follow the line. That was the only instruction we had, "follow the line."

The line went up to the second floor and once we got to the second floor we had about 50 guys in a smoke filled room that could hold about 50 guys, but by the time we reached the 2nd floor, we had run out of air. The instructors said there were no further instructions and there was nobody guiding us up the stairs, in fact there were 2 or 3 metal pans in the middle of the floor filled with burning tires and straw. The smoke was as thick and black as it could be. I followed the wall in my search and I had I guy who had never been in the fire service before, Kevin Shay, he was sort of hanging with me and guys around us were beginning to panic. We could hear grown men starting to panic. I was out of air myself, and the procedure back then was to take your low pressure hose from your mask and put it in your armpit and hopefully that would filter out the carbon or some of the bad stuff. You would still be breathing whatever air you could find, but this included carbon monoxide, sulfur dioxides, all the bad stuff, you are still breathing it, but this method helps to filter some of it out.

As I was following the wall, listening to people panic, I had to stop myself and say, "Wait, wait, wait, this is crazy." Grown men are getting up and running at full speed across the floor, slamming into walls, other people were starting to lose control themselves, and I had to stop myself. This guy Shay is with me and he is asking me, "What do we do, what do we do?" At that time that was a good question because it made me stop myself and think, if I were back in Richmond, in a burning building, didn't have any air, what would I do? Get to a window. The windows on this burn building had big iron shutters so as I continued to crawl along the walls, I reached my hand up along the walls until I felt one of the big iron shutters and I proceeded to open the window. Our lead drill instructor, Lieutenant Sherman Powelson, happened to be standing on the ground when I opened this window. He looked at me and called me everything but a man. It turned out that this was a panic drill and the reason for the drill was to put us in an uncontrolled environment with no instructions, no leadership, just to see how we behaved.

Essentially (I didn't know this at the time) when I opened the window, the drill was over. Somebody had taken the leadership role, somebody had acted on trying to improve the situation instead of these guys on the second floor calling for their mama and crying.

In the meeting after the drill, all the white firefighters in the class looked like we were in minstrel show. The smoke was that black and thick, even with our face pieces on, breathing through our low pressure hose in your armpit. There was enough carbon in the air that all that could be seen was the whites of our eyes and our teeth. The carbon just made the black firefighters blacker. A big discussion ensued afterwards about how those who panicked and were not sure of themselves or didn't know what to do, then maybe those folks should think about doing something else for a living because the fire service is not for them. As I said, that was the only time I have ever been involved in something like and it was very interesting. That summer I learned a lot about myself and I am sure that everyone else learned about themselves too.

Because we were getting beat up so hard in class and because the training was tough, several of us went out in the evenings and did what grown men do, had a few beers and some fun. There was one guy in the class, there is one in every group, I guess that is sort of a little behind socially or whatever you want to call it, but we had a guy that I'll call John. He was raised by his grandmother and a good guy, he is an officer in D.C. today, but John was a perfect target for any kind of tricks and puns and ploys. John lived by himself in his own apartment and drove his grandmother's car. He was a good guy, but he did catch a lot of grief. Anyway we were out drinking one night, it was probably around 11:00 and we had to work the next day, but we decided to give John a visit. We rolled up on a 3 story garden apartment project and John lived on the second floor. As we ascend the stairs, Zeke decided to take the 2 ½ gallon pressurized water can fire extinguisher from the stairwell. I knew this was going to be funny.

We knocked on the door and of course it is way past John's bedtime and there were four drunken firemen coming up the stairs, making all kinds of noise to the point where the neighbors are coming to their doors. Well, John makes the mistake of cracking his door open, he had one of those little burglar chains on, which when Dipski kicked the door broke and the trim and frame of the door went with it. Then we bum rushed John's apartment. Zeke was running around in circles through the living room and dining room which were connected in a circular layout, discharging the fire extinguisher against the ceiling which in turn, caused the ceiling to rain down dripping on us: just drunken nonsense. We raided his refrigerator, I think he had some Pabst Blue Ribbon beer that we confiscated and some grapes, then we abruptly left as quickly as we appeared, leaving poor John wide awake, cursing and fussing at us, with water dripping from his ceiling.

We got to work the next day and although it was all in good fun, John was some kind of upset and informed us that the D.C. police had come the night before and taken a burglary report and he was going to have all of us arrested. It made for a couple of interesting days as we waited for the police to show up to lock us up, which they never did. I imagine, knowing John, and all the crime and trouble in the District, the policeman probably got as big a kick out of it as we did. Poor John, he was always the butt of every joke.

The ten weeks flew by and during the last week of class we received our assignments regarding where we would be stationed. Now, I was about 23 years old and probably spoiled because I had been assigned to the busiest Engine in Richmond and was used to going to tons of fires and really enjoyed it. I learned a lot in the D.C. academy that made me a better fireman. There were 32 firehouses in the District of Columbia, and I could think of only 3 that I did not want to go to. Lo and behold when the assignments came out, I was sent to 14 Engine, #1 Platoon, which was one of the three I didn't want to go to. I took that as an omen, a sign from God, that maybe I didn't belong up there. But, I had come this far, endured 10 weeks of training, and was graduating at about 160 pounds and in the best shape of my life. I had 49 classmates, 45 of which were like best friends to me. I couldn't have found a better group of men. It was a great summer all the way around. So I figured I would give 14 Engine a try.

Now Richmond had a policy where I had a year during which time I could return to Richmond and be reinstated, with no questions asked. All I had to do was ask to get my job back and since I left Richmond under good terms this was in my favor. D.C. back then had 3 shifts; you work three 10-hour days, three 14 hour nights and then get a 3 day break. My shift as it landed began with my first shift being on night work. I met my captain, which was already the start of something bad. This guy's name was Ronnie Taylor and he was an old-school D.C. fireman, probably not a bad guy, but I was 23 years old and full of piss and vinegar and really didn't want to be at that station, and wasn't in the mood for no dumb shit so to speak.

Captain Taylor and I got off to a rough start because of my bad attitude. He sat me down, and I didn't want to lie to the man, but he wanted to know if I had any experience. I really wanted to start off fresh in D.C.; I didn't want to be carrying any baggage from Richmond, but I didn't want to lie, so I told him I had 3 years in Richmond. His first comment to me was, "Good, you got the line, if we catch anything, just hold on to my running coat." Well, I knew we were off to a bad start from the get go. This guy was 60 pounds overweight, probably in his mid-forties and here I was in the best shape of my life. I could run circles around this cat…still. He had quite a large ego which was evident and as reputations go, 14 Engine did not have the best reputation in the fire department either which didn't help my attitude any. Although there was an ambulance and medic unit that ran out of the house, we were still considered a single house. We came on duty at 1800 hours and it was customary for the officer to sit the first watch from 1800 until 2200 hours. It's a four-hour watch, but I don't think that any officer actually sat at the watch desk for that entire time; he would look out for the phones, that sort of thing. Also, it was customary for the rookie to have the 0200-0400 watch which was a time when everyone was in bed, sound asleep. The watch desk was a duty where you had to stay awake and make sure that everything was ok on the apparatus floor, monitor the radio, answer the telephones and answer the front door. It was like the nerve center of the firehouse, the reason being that at 2:30 or 3:00 in the morning it was the watchman that turned on the bedroom lights, rang the gong, and copied the address and turned out the company for a run.

Occasionally, we would get someone coming to the door to report an emergency, so it is a good idea to have someone manning the watch desk during the night. Captain Taylor told me he expected me to sit at the desk from 6:00 p.m. until 10:00 p.m. and study my books, which wasn't a bad idea because there is a lot of stuff to learn in that first year, including box numbers, the streets, the running routes and stuff like that. On top of all this I was never really able to sleep before 1 or 2 a.m., it was my custom in Richmond to stay awake until one or 2:00 a.m. because in Jackson Ward, things didn't really come alive until 9:00 p.m. and therefore I wasn't used to going to bed before 1 or 2 a.m. at the firehouse.

Now, **my** watch was to begin at 2 a.m. but because it was customary for each man on watch to sit an extra half hour so the man for last watch wouldn't have to get up, my watch actually began around 3:00 and lasted until 5:30 a.m., when the day shift began arriving to work; it looked like I was going to be up all night just about every night on night work Night work to me meant no sleep, no sleep at all.

The thing that did bother me was when I sat down and went through the entire log book for the previous two years, I couldn't find but a couple of working fires. The company ran about 1500 runs a year, but the calls were all medical assist calls and auto accidents, very few fires. Once again, I was disappointed. The other thing that really sucked was that every night at the watch, I could listen to the city radio and hear Engine 25 would have a fire, Engine 6 would have a fire, Engine 12 would have a fire, and we would just sit in quarters. This only increased my feelings of disappointment. After being in the engine house about a month, I pretty much decided I was going back to Richmond.

I came to work one afternoon around 3:30 P.M. (the shift didn't start until 6:00, but it was a man for man deal and as soon as I got in, the man I was relieving could go) anyway, on this particular night, the man I was relieving had the layout, in other words, he was taking the hydrant. I had just placed my gear on the wagon after being in the firehouse for less than 10 minutes when we picked up a second due street assignment. It was Engine 22, 14, Truck 11, Chief 4, for smoke in a building on 5th Street NW, 5th and Missouri, NW. It was about 4:00 in the afternoon and Missouri Avenue is a pretty big commuter route added to the fact that traffic is pretty bad in D.C. almost any time of day, we responded 2nd due on the assignment.

According to the Standard Operating Procedures, the alley to the rear of the building would be our responsibility. We arrived at 5th and Missouri and as we came up to the alley from 4th Street, we were stopped in the middle of all this traffic and the captain was yelling at me to lay out. When I got down to look, we had no access to the alley so I looked at the wagon driver and he motioned to get back on, and before I could even do that, he made a right turn onto 5th and took off in the opposite direction of the fire building. So there I am standing in the middle of Missouri Avenue with no fire truck. At the time, every engine company in D.C. was a 2-piece engine and the pumper driver, Marty Fulstein, looked at me and motioned for me to get the backstep of the pumper. Everyone else is on the wagon and me and Marty are on the pumper and we make a big circle around the block until we locate the alley that we are supposed to be in. When we get to the alley, we were going to split lay; the wagon was going to lay from the street up to the alley to the rear of the building and the pumper was going to lay from the alley up the street to the hydrant, A split lay. Since I was on the backstep of the pumper, Captain Taylor doesn't' know where I am, so now I am out of position. We finally catch the wagon and the lineman had gotten off and laid out and the wagon took off up the street with the lineman standing there to complete the layout, leaving the captain and the wagon driver on the wagon. The more this goes on, the more it resembles a bad episode of the Keystone Cops or the Three Stooges. We arrived on the pumper to complete the split lay and I step off the back step with the lay out section and motion for the pumper driver to go as I am looking at the lineman. The lineman looked at me as if to say," Where were you at?" I told him I would make the hook up and he can go the heck up the street.

He goes up there to run the line (keep in mind the rear of these buildings in D.C. are fortified). By the time he runs up the alley, pulls the inch and a half attack line and gets to the rear of the building, the captain is back there and has discovered that he can't gain entry because the doors are all barred up. After I hook the 2 layout sections together and flake out the hose, I run up the alley, grab my breathing apparatus, and get accosted by the captain, whose face was beet red, was yelling and screaming to get a Halligan bar to force the door. I do as ordered and get up in there; meanwhile 22 and 11 were already coming in from the front of the building (this turned out to be a small fire in the walls and the basement). We gained entry just in time for the captain to tell us to take up. We needed to go back in service, pick up our hose and go back home. That was my first rodeo with Ronnie the clown.

After this fiasco, I had firmly made up my mind to return to Richmond. The very next night, the circus was still in town. This story illustrates why I really had to go back to Richmond.

I came in to work (the second night of night work) and at about 9:30-10:00 in the evening, we get dispatched 1st due for a box alarm. A box alarm in D.C. is a big deal. In fact some of the guys who are normally calm, cool and collected go from Jekyll to Hyde and become unglued when a box alarm comes in. Now our wagon driver wasn't one of these guys, he was one of the nicest guys you would want to meet, but first due on the box carried a sense of urgency. We were first due and by God we were going to get there first and God help anybody who might get in our way. I was running the line that night (the box was at the Soldier's and Airman's Home off of Shepard Street NW) and as Sam drove the wagon, he had it running as fast as it would run and was coming through the gates of this semi-private almost estate-type deal, we arrived first in side 1 on a 4-story retirement home.

The front of this particular building was a 4-story glass wall, with a foyer inside the entrance. You could see through the glass that there was smoke on the 2nd floor. The hallway runs from the entrance foyer directly to the rear of the building. Now you can see smoke coming toward the front of the building from the rear.

The captain yells for me to pull the 400, which I promptly did. Now the 400 is a tricky line that takes 3 men to pull. The nozzleman takes a hundred feet, shoulder load, steps off to the side then the officer takes a hundred feet does the same thing and the wagon driver takes a shoulder load of a hundred feet, reaches back and pulls another hundred feet and drops it on the ground. As you walk away from the fire truck toward the objective, the wagon driver holds his stuff on his shoulder tight and stretches out as much on the ground as he can and flakes it out, then he plays off the hundred feet on his shoulder. When that is complete, he drops out of the line and heads back to the wagon. Next, the officer plays off the 100 feet of his shoulder and finally the lineman plays out the last 100 feet. It is a little bit of a tricky line to run, but not too difficult if you take your time and communicate and work together. It is obvious that we have some kind of fire in this building and the captain is once again excited and his face is all red and he is screaming in the radio. So I took my hundred feet and stepped aside, the captain gets a hundred steps aside; Sam gets his hundred feet reaches back and pulls the last hundred to the ground. We make it no more than 25' from the wagon when the captain takes his hundred feet shoulder load and just dumps it outside the front door and runs up the stairs towards the smoke. Sam and I made eye contact and as we looked at each other we were both thinking, "Can you believe this crap?" The captain just took a line that is hard to stretch under good conditions and damn near made it impossible now. In the meantime, Sam sets his hundred feet down and motions for me to keep going as he tries to untangle the mess of hose on the ground that the captain just dropped. I start in the front door (remember there is a 4-story atrium with a staircase that sweeps to the right) I have 100' of inch and a half on my shoulder and had to pull the hose that I was dragging over the banister so I could stretch more directly than if the hose went up the steps. I figured this was the best shot we had to clean up the mess the captain just created. In the meantime, the captain ran down the 2nd floor hallway and disappeared into the smoke, screaming on the radio the whole time. Working together, Sam and I try to make chicken salad out of this chicken shit the captain just created. I pulled the hose down the 2nd floor hallway as far as I could with Sam feeding me as much line from the ground as he could. I made it about 100' down the hallway and with the weight of the hose coming over the railing I couldn't pull any more from ground level so I began to lay from my 100' shoulder pack. I yelled down to Sam that I thought we had enough hose and then proceeded down the hallway to the fire room. I get back to the doorway of the room that is on fire, and I am literally 10' short. Ten feet of hose shy. I looked around and in all this confusion and aggravation I didn't have time to put my face piece on, smoke is down to about knee-cap level, and the captain is standing in the hallway screaming for me to go back and get more hose. He snatched the nozzle out of my hand and screamed for me to get more hose. All we needed was 10 feet.

While this was going on, the very next door beyond the fire room is a staircase in the rear of the building. I don't know if you have ever heard firemen stretching a hose up a staircase with their air bottles on (and this is back in the day where the air bottles were made of steel) it sounded like bulls in a china shop.

We could hear Engine 24 coming and they were coming hard. So now company pride is on the line. Remember I had pretty much decided to quit this fire department and I am disgusted and not real impressed with this bozo captain we got who wants me to hold onto his running coat. I looked up and saw a 2 ½ gallon pressurized water extinguisher on the wall. I look into the room and the fire appeared to be confined to the mattress and the curtains and one overstuffed chair. Instead of following the captain's orders, I snatched the can off the wall and proceeded to knock the flames down. By this time, I was so damn mad I couldn't see straight because the 400 line is never an easy pull, but it is impossible when you've got a 40-something year old veteran bozo screaming and hollering on the radio because there was a little bit of smoke in the building. I never understood these guys who got to screaming. We are professional firemen. The building is on fire that is why they called you, fix the problem; don't hoot and holler and scream. I never understood those guys and never had any patience with them. In a fit of anger after I knocked these flames down, I snatched the mattress up and folded in half, and was eye-to-eye with this captain and I said, "Here Captain, what do you want to do with this?" in a real deep angry voice. He looked me in the eyes and said, "Take it outside." As I left the room I passed the guys from 24 and was carrying the smoldering and flaming mattress, and made my way down the length of the building, to the front of the building and out the front door with the mattress. Of course when I got outside, I told Sam what had transpired and how this idiot wanted me to run 200' back up the hallway to get 10 more feet of hose, as we could hear 24 closing in for the kill.

Sam and I were both pleased that I did not allow 24 to get the fire. Even in an outfit like Engine 14 in D.C. you don't want anyone walking over top of you. It's a pride thing. The thing that made the rest of the tour interesting was that I could tell that the captain was proud of me to for knocking down the fire and carrying the mattress outside before 24 got there, but he was also pissed off that I didn't go back for that 10' of hose he wanted me to get. I think he realized that he dropped the ball and he knew that WE knew that he dropped the ball and he knew that his rookie was not impressed with him at all.

It made for an interesting night back at the firehouse after this was over. The troops, the firefighters, at 14 didn't have a lot of respect for this guy, and it was evident that he was a captain in name only. The very next night of night work, the third night, I made a mistake that could have cost me my job on the spot. With the tense situation that was going on, it was the wrong time for this to happen.

I came to work that night about 3:30 p.m. already tired, I think I'd been up for a day and a half by that point. The evening progressed without anything exciting happening, and at 2 a.m. I found myself sitting at watch desk on the apparatus floor between the two overhead doors. This was probably early November and the leaves on the ground were blowing down the street, I got up a couple times to look out the window at the Rock Creek Cemetery which was across the street, just trying to do anything to stay awake.

At the watch desk at every D.C. firehouse is a leather-bound log book with numbered pages; in essence it is a legal document of anything that happens at that fire station. Every shift logs in this book. This book sometimes lasts between 6 months and a year depending on how busy the station is. You could refer to this log book and see the events that transpired while you were on a 3-day break to get up to speed. The runs (any emergency calls, service calls) were entered in red ink, with the box number, incident number, what the company responding to, the time the call was dispatched and the time the company returned to service.

Any regular business such as the captain or lieutenant on duty, wagon driver, pumper driver, layout, small lineman, and those on a regular day off was entered in blue ink. As the guys came on duty, the first thing they would do is remove the gear of the man they were relieving from the wagon, put his gear in its place and check his air bottle real quick to make sure it was ready to go, he would then go to the desk and sign the book. For example, if I were relieving the lineman, I would sign in the book where it said lineman and put the time in there and the air bottle pressure from the Scott Pac. From that moment on, I was legally responsible to respond to emergencies and the person's gear I had removed and hung up on the gear rack, was now relieved of duty and could go home.

As I said, I was totally exhausted and trying to do anything to keep awake when I awaken to the telephone ringing at about 5:30 a.m. and found that I was face-down in the log-book and during my slumber, noticed that I had drooled in the book. There was a pool of my slobber lying on the two pages of this legal document on which was recorded the past two days of activity. In state of panic to dry the book as quickly as possible, I used my bare hands to try to wipe up the mess, but ended up smearing the ink to the point where it was almost illegible. As I looked at the slobber saturated book, I knew that when the book was seen, it would be obvious that I had been asleep, and was an offense that you could be fired for: sleeping on the watch.

Luckily, one of the dispatchers that night was a volunteer from P.G. County that I knew pretty well named Buddha Burke and I quickly picked up the phone because my next thought was what if we missed a run? I told Buddha, "Buddha, this is Jake at 14 Engine, we haven't had any runs in the past 3 ½ hours, have we?" Buddha responded laughing, "You've been asleep on the watch, haven't you?"

I said, "Yeah, Buddha, I was fighting it and I passed out, man I am scared to death." Buddha was still laughing and told me, "Don't worry, kid, I got your back. If I didn't get the acknowledgment light from you, I would have rung the phone right away and called you. These kinds of things happen in the city from time to time. But we're not going to let you miss a run." That was my 3rd night of night work.

When I got off that morning, I was already dead tired, I ended up sleeping in the bedroom until noon, at which point I got up, got in my car and drove straight home to Richmond, where my wife and son were still living. When I got home, there was a phone call from the fire chief's office and the very first thing I did was to make an appointment to see Chief Lewis in Richmond, so I could see about getting my job back. With the appointment made for the next day I went down dressed in a suit and tie to meet with the Fire Chief of Richmond.

Chief Lewis was very cordial and asked me how things were in D.C. and what didn't I like about the job, and this, that and the other. I explained to him that I was really miserable being assigned to 14 Engine and there were a lot of things about the D.C. fire department that were not as I pictured them in my mind to be. I admitted that the grass seems greener on the other side and I didn't know how good I had it in Richmond. He assured me that he would take me back, but he couldn't guarantee that I would be stationed at 5 Engine, he didn't know where the vacancies were at that time and the meeting concluded. During my stint in D.C., another recruit class had gone through and my former position at 5 had indeed been filled. I was happy as I left the Chief's office but still a bit apprehensive knowing that I could be leaving a slow firehouse in D.C. only to be assigned to a slow house in Richmond. I had made the move, so I knew that regardless of what happened, I had nobody to blame but myself.

When I returned to D.C. for day work, I had planned to give my 2 week notice, but I wanted to have a little fun with this captain first. It didn't take long. When the captain came in, and the first words out of his mouth were that we were going to go out and practice pulling the 400' attack line around the firehouse today. This was met with snickers and guffaws from some of the men because we all knew that the reason the 400' came up short at the Soldier's and Airmen's Home and he knew it too. It seemed like a dumb thing to even talk about unless he was practicing for his own benefit, since the rest of us had it down pretty good. When I finished recruit school in D.C something I (and the other recruits) could do in our sleep, is pull attack lines, throw ladders, and things of that nature. Well, we went out and pulled the 400 about six times, and needless to say, the character of this fellow was exactly as I thought, each time we pulled it, it was left for me, the wagon driver, and the pumper driver and the layout man to pick up all the hose and fold it neatly and properly in the hose bed. We finished the drill, and our captain retired to his room to do his paperwork.

One thing that did happen during this period of time was that I had become really close to the wagon and pumper driver and the guy that usually ran layout and it was pretty much unanimous that nobody in the house could stand this captain. The firehouse had a nickname given from other firehouses, they called 14 "The House of No," no eating fund, no commissary, no this, no that, no firemen, and I could add no fun to that.

Interestingly enough, in the kitchen, where what would have been the commissary, was a little homemade sign made me chuckle every time I looked at it. It read, "Danger is no stranger to a Fort Totten Ranger." Fort Totten was the area where 14 Engine was located. Danger might not be a stranger to a real fireman, but it didn't seem like these guys went to too many fires and this along with the leadership and some of the characters that worked at this station just left me scratching my head. It was nothing like I had ever seen before during my 3 ½ years in Richmond.

The second day of day work, the entertainment continued. I was still looking for a way to give my 2 weeks notice and was looking forward to messing with this captain, no holds barred, until he tried to fire me or write me up on charges. I had nothing to lose, since I already had a starting date to return to my job in Richmond. The captain took the company to a Special Education School named Mamie D. Lee which was located on Fort Totten Drive about 5-6 blocks from the firehouse.

We went there and found numerous safety violations; ridiculous stuff that the D.C. government is famous for, it is amazing that some of the buildings in D.C. are not just falling down from lack of maintenance. As we waited in the parking lot, after the inspection, for our illustrious captain to finish his talk with the principal of this school, Sam looked at the pumper driver and said, "Yeah, when we pulled in here the captain had a tear in his eye." Of course the rest of us were waiting for the punch line and he continued, "It's always hard to come back to your old alma mater."

    I was still waiting for my opportunity to get this captain fired up, but didn't want to make it obvious so I decided to up the ante a bit on the first night back on night work. We came back the 1st night, and I had been instructed to take the additional watch the first watch which was usually the officer's watch from 6-10 p.m. and use the time to learn the rules. I felt that I had a good handle on the area since I spent my time off riding around in my car trying to learn the area. A good rookie does this without being told. My friends, Zeke and Dipski, had an apartment together and they would let me crash there during day work. I used to leave Richmond at about 4:00 a.m. on my first day of day work, and then stay in D.C. until my 3rd day, at which time I would drive home to Richmond to spend 24 hours and return to D.C. to begin night work. My first night back at night work, about 8:30-9:00 p.m., I wasn't really tired, but had to mess with this guy and now was my chance. I went into the bedroom and turned the lights off and got into bed. Now during this time of day at the firehouse, there are plenty of people wandering around the firehouse, on the apparatus floor. The phone gets answered, so you are not going to miss a run since the speakers are still on throughout the firehouse. The speakers don't get turned off until the first person goes to bed, but I left the speakers on. I told Sam the wagon driver to watch this, as I upped the ante. It wasn't long before the captain discovered that his new problem child was not sitting in the place where he had been told to sit a couple months ago. Keep in mind that I had been busting on this captain when the opportunity presented itself. For example, I did comment to him, "nice job on the 400, captain" after we had saved the day with a 2 ½ gallon fire extinguisher. So, there I was, laying in bed and about 9:15 p.m. I hear a pair of heavy footsteps stomping across the floor. I was so excited I couldn't sleep, so I peeked through one cracked eye and saw the captain who had come looking for his rookie. Well he knew where my bed was, and he came over and looked at me. I was waiting for him to say something, but I don't think he knew what to say. He proceeded to walk out the opposite door and in about 15 minutes he returned and this time, simply walked through the room, without pausing at my bed and left through the opposite door, as if he was pacing in circles. At 10:00 p.m. when my (extra assigned) watch would have been up, he came in, turned the lights on, and by this time Robbie and the pumper driver had gone to bed, and the captain walks directly over to me and puts his finger directly to my nose in a confrontational way and said, "Didn't I tell you that you would sit the first watch every night, rookie?" I feigned waking up from a sound sleep, pushed his finger out of my face and replied, "Yeah, captain, I seem to remember a conversation like that." He said, "Well then, what were you doing in bed?"

I replied, "The rule book says that from 6:00 until 10:00 is your watch, and since you are a stickler for the rules…" At this point he erupted and naturally so. I would have probably done the same thing to a rookie. He told me to put my clothes on and report to his office forthwith. By this time the other 3 men at 14 Engine are snickering under their breath, and were wondering what this crazy ass rookie has up his sleeve. I knocked on the office door, and was admitted where I was given a butt-chewing that lasted about 20 minutes. His comments included, "You don't know who is in charge in this company…I gave you an order some 3 months ago and you are disrespecting me and violating my orders…."

I sat there and took it for a while, and then calmly asked to see the chief. The captain asked me why and I replied, "You and I can't seem to get along, I want to request a transfer out of this hell hole." At this point, the captain lost all self-control and threw me out of his office as he told me to go sit at the watch desk. I went over to the watch desk about 25' from his office and I sat there for about 5 minutes at which time he orders me back into his office. I guess he needed a few minutes to regain his composure. It is now about 10:30 at night now, and you don't disturb the Battalion Chief in the D.C.F.D. after he's made his rounds, unless there is something very important going on, or if there is a fire emergency.

Captain Taylor picked up the receiver of the rotary phone, and as he dialed, I thought that he was going to spin the dial right off the phone base. His fat fingers were turning the dial so hard; I just knew he was going to break it.

When he got Chief Rider (who I would like to apologize to at this time, since I don't think he knew what was going on) on the phone, he was practically screaming at the Chief that he needed him to come to 14 Engine because the rookie doesn't seem to understand who is in charge here. I was dismissed from the office and sent back to the watch desk to wait for the chief to arrive while the captain finished the phone call. I sat for about 20 minutes until the chief arrived.

His driver was a real nice guy named John. John was walking behind the chief with a grin on his face, looking at me and then he grabbed Sam, the wagon driver, and headed to the back corner of the apparatus floor to get filled in on what was going on. That was typical of the D.C.F.D. in that most times, the aides knew more of what was going on than anyone else. It was also the aide's job to keep his finger on the pulse of each engine house as well as payroll and other administrative duties. If the chief wanted to know what was going on, he would go to the aide. After Chief Rider and Captain Taylor talked for a few minutes, the chief launched into a 10 minute butt-chewing session which he concluded with, "You know you are still on probation and I could have your badge right now, tonight." I replied, "Yes, sir, I understand that. Can I say…?" At that point he interrupted me and said, "No, you can't say anything. If this man tells you to climb that flagpole out front, at midnight tonight in your underwear, I expect you to be on your way up the flagpole. Do you understand me, son?" I said, "Yes, sir. I understand."

I was then dismissed. At this point I looked at the captain and asked if I could go to bed now, which just pissed him off even further.

On my way out, I was waylaid by John and Sam who wanted to know what I was up to and I gave them the Reader's Digest version that I wanted to be transferred out of 14 and had asked everyone about this already and was told that I had to serve at 14 where I had been appointed. A few minutes later, the chief comes out and of course John hurried to drive him away and the chief added a comment as he crossed the apparatus floor, "And I can tell you something else, young man, if I ever have to get out of bed in the middle of the night to come up here and deal with something like this again, it will be the last time. Right here, now is the last time." To this I saluted him and said, "Aye-Aye, sir." That was night number one.

The next morning when my relief man got there, I was trying to get the heck out of dodge, the captain caught me and warned me, "When you come to work tonight, son, you be ready to work." Okay, no problem. I got there the next night and discovered what ready-to-work means. The fire house was filthy. There were no cleaning supplies, window panes were broken, and hadn't been fixed in years. Pretty much what needed to be cleaned like the kitchen, the soap and cleaning supplies were purchased by the men themselves. This was during the Marion Barry administration. It seemed like everything was broken.

The captain went to his car and brought in soap powder, brillo pads, everything you could imagine as far as cleaning supplies. The stove and the refrigerator were next to each other and both were white in color and looked like it had been years since either one had been cleaned. My job that night was to clean the refrigerator of which the left side was coated with baked on grease from the stove. I pulled the refrigerator out and proceeded to scrub the exterior with a brillo pad and soap and water. After about 3 hours, I had it looking pretty good. The captain told me to let him know when I was finished. He came out and seemed pleased the refrigerator was actually white again.

He then went out to the back alley and returned with a can of automobile wax, and wants me to put a good coat of wax on the refrigerator. Well, I have waxed a lot of things in my life, but I have never waxed a major appliance. I was like, touché, he had one-upped me. I finished waxing the refrigerator and it was about 11:30 at night and the refrigerator was looking good, I have to admit. When I was in high school, I had an automobile that I waxed twice a week, and one of the things that I used to do back then to seal the wax on was to run clear, cold water over it to harden the wax up and it would really look shiny. Since the captain had one-upped me, in my smart-assed brain, I got to thinking, "Hey, anything I have ever waxed, I have always run a coat of clear cold water over to get the shine." I didn't know if we had a hand truck/dolly so I got Sam and a couple of other guys to help me find one and then we removed all the contents to prevent breakage, and moved this freshly waxed refrigerator onto the front ramp where I wet it down with a garden hose.

True to form, Captain Taylor came out of his office, looked at me and asked what I was doing. I explained to him that the instructions on the wax tell you to run a clear coat of water over it to harden the wax. The captain spun on one heel, returned to his room with a look on his face that I was the biggest dummy in the world, which pleased me very much since I got the reaction I was after. After I dried the refrigerator off, we trucked it back into the kitchen and loaded it back up. Sometime while we were loading the refrigerator up, the captain came in and let me know in no uncertain terms that I was expected to be ready to work at any night at night work. Back in those days, all the housework and building inspections were done during day work. Night work, since you were away from your home overnight, was basically answering alarms and making sure the fire equipment was ready. Night work was in essence just showing up and being there to answer fire calls. The captain let me know, that until he saw fit, every night of night work I was expected to be ready to work.

The third night of night work, I am still looking for my grand exit; something I could do to have these guys talking about this crazy rookie from Richmond for years to come. The chief's house is over at 11 Engine at 14th and Newton Streets N.W. It housed 11 Engine, 6 Truck and the 4th Battalion Chief. On that afternoon at about 5:00, even though we weren't officially on duty, we fired up the wagon and the pumper and headed over there. The chief had said that they wanted to see what I knew and had set up a practice session. By now, both the chief and the captain had it in their heads that I was a smartass and hardhead and now they wanted to see if this young whipper snapper knows what he is doing. They were looking for any mistake I made so they could scream at me in front of my peers. They wanted to bring the young guy into the fold and take the boldness out of him…maybe he is not at hot as he thinks he is. All this was to be done in front of about 15 firefighters who were senior to me. The goal was to show me my place as the one at the bottom of the food chain. After the chief explained what he wanted us to do, with the help of a couple of guys from 6 Truck, I threw a 35' ladder to the firehouse roof at the rear of the building and went up and tied it in using the rope hose tool. The D.C.F.D. had a technique to do everything. The chief explained to the guys from 6 Truck that he wanted me to call all the shots on how we were going to position this ladder and where we would put it and they were there only to assist me on raising it, since a 35 is about 200 pounds or so. D.C.'s policy said that a 35 was a 3 man raise or a 4 man raise; I really don't remember, but we were to do this according to D.C.F.D. procedure. After I tied the ladder off and came back down the ladder, the chief went up himself and checked the knot. When he didn't say a word, I knew I had it because any mistake would have resulted in a brow-beating session. The chief told me to go back up the ladder, untie the knot, bring down the rope and put the ladder back on the truck. He then told me there was a fire on the 2nd floor of the building and I was to use a particular stairwell and he wanted me to run the 250 to the 2nd floor.

The stairwell at Engine 11, top and bottom has automatic door closers which automatically closes the door after you enter the building using a hydraulic arm. Now one of the most basic things an engine man knows is that if you run a dry hose line under a door like that as soon as the line fills up with water, it is going to hit the space where the hose was under the door which is about 3/8 of an inch and the water is going to get stopped right there and it would be difficult to get back out to open the door; it becomes like a chock at the bottom of the door.

I stretched the 250 which is a one-man stretch, 100' on my shoulder, reach back and drop the other 150' on the ground and by the time I got to the door I had about the first 150 on the ground flaked out since the staircase was at the opposite end of the building. I reached up to my helmet where I always keep a couple of 8-penny nails in a rubber band and put a nail in the door jamb so when the door closed it would close on the nail and the door would be held about ¾ of the way open. The Battalion Chief was observing the whole process and when I arrived at the top of the stairs, before I opened the door to the bedroom, I asked him exactly where the fire was supposed to be on the second floor and what the smoke conditions were. He asked me why I wanted to know and I replied that I had about 75' of hose on my shoulder and if the fire is on the other side of this door, I will need to call someone to flake this thing out for me or I need to do it myself. Secondly, the hose is still dry and if the fire is inside the doorway, I was not going anywhere until I had some water. This seemed to please the chief. At this point the chief could tell that the problem may not be totally with the rookie, but that perhaps the officer might be a knuckle head too. I am sure knowing how fire departments are, that he did have some idea from stories what Captain Taylor was all about and that he just had a bad mix here; a young aggressive new fire fighter and an old redneck hothead captain.

Chief was quite pleased and told me I had done good and to go and rack the line up, so I took the 250 back down the stairs, and he asked me why I had chocked the door and I explained to him the reasons and after a few more questions about ladders, he told me, "That's all for today, kid, you done good."

Organizationally as far as the fire department goes, he couldn't say much more to me because once again I was the one rocking the boat. Here is the best part of that story, after we got the hose back on and were getting ready to leave, I asked the captain if I could speak with the chief in private for just a minute. The captain wanted to know what I wanted to talk about with the chief and I told him it was kind of private. So he said no, but he would go ask him. He came back down stairs and told me I could speak with the chief but he (the captain) had to be in there with me. I told that captain that that defeated the whole purpose of a private conversation and that he knew that what I wanted to talk about was him and that he was the reason I wanted to speak with the chief. This only pissed him off. By now the chief was involved and upstairs we went, we began by me asking the chief if there was any way I could get a transfer. I told him I love the job, I just can't work with this man right here. Captain Taylor was standing there listening and looked as though someone was poking him in the ass with a hot poker. The chief replied that he had already explained to me that I would have to work my first year at 14 and that is just the way it is. You have to just suck it up and do what the man says and try to get along. While he was speaking I was undoing my badge and then taking it from my shirt told him, "I am not going to work here anymore, I quit." Right then the atmosphere changed, it seemed as though the oxygen had been sucked out of the room. Here's a captain who had been at war with me for 2 weeks, now suddenly concerned that I had a young child at home to take care of. I looked at the captain and said, "I don't care if I have to pump gas for a living, I am not going to work for you, and don't you understand?" The chief shoved the badge across the desk and said, "Son, can you take a day or two to think about this? I think this is rash decision, I don't think this is in your best interest. Everything that has happened up to this point has been positive. I have heard nothing but good things about you." Both of them were trying to talk me out of quitting. I said, "No, sir, I'm done. I can't do this any more." So the chief sits back and thinks for a minute. He then says, "Do you think you can do one thing for me, son?" Now remember that famous rule book? One of the rules in the D.C. rule book was as soon as you submitted your resignation, you were to take all your gear off the wagon and you weren't to answer any more emergency calls. The thought process behind that rule being that you couldn't get injured on purpose or fake an injury to get workman's comp. That made perfect sense to me. So, back to the chief's question. I replied, "Sure chief, what is it?" He asked me if I would finish working the night so he didn't have to hire an overtime man. I told him I would.

When we got downstairs, it was apparent that the men knew that the rookie had just resigned. In fact, one of the men was probably upstairs with his ear to the door and reported back to the others. A few men tried to pull me aside and talk me out of it. Zeke's captain at Engine Co. 27 got wind of what had happened and called me to try to talk me out of it. It was kind of comical.

The cruel thing on my part of it was that I didn't tell any of them that I had gotten back my job in Richmond. There were a lot of little things that I saw as omens; none of which were going in my favor. I felt that it was time to cut my losses and go back to Richmond. The best part of this experience was that I got to meet a couple hundred great, great firemen; I went to school for a whole summer and learned a bunch of new tricks that have helped me since. The icing on the cake here was when I got back to Richmond when I was to report to training school and pick up my turn out gear; I had a letter waiting for me at training that assigned me back to 5 Engine, B Shift. I was going back to 5 Engine after a couple of month's absence with the same locker, same bed, and same badge number. It was almost as if I hadn't left. But it was the best summer of my life.

- Chapter 15 -

Back in Richmond

On my first tour of duty after I returned to work in Richmond, I found a new rookie there by the name of Mikey Edmonds. Mike had just graduated from Longwood University with a bachelor's degree and he was a clean-cut white guy, same age as me who wore his hair just like Mike Ditka, the coach of the Chicago Bears. Mike had never been in the fire service before and the shift at that time had really changed during the time I was gone. It had gone from a shift with mostly older guys to one with young and I hoped soon to be aggressive guys.

I knew it would take a few fires before we could get into the groove and learn how to work as a team. After a while, we became so good at anticipating what each other would do, that we would go to a fire and didn't even have to speak to each other…we were running like a well oiled machine. One of my first calls after coming back from D.C. was a fraternity house on W. Franklin Street at VCU. This fire occurred on about a 15 degree night; it was about 1 in the morning when the box opened up. The assignment was 10, 6, 5, 33, 43, 41, 52, for fire on the 3rd floor of a brick and joist building. The wind was blowing pretty good from the south on this night and when we pulled up the 3rd floor windows had failed and we could see where fire had come out of one of the windows earlier. It seemed strange, the trim around the window was burned pretty good, but there weren't a lot of flames coming out of the window. What we didn't know at that time was that due to the southern wind, the fire was being fanned by the wind coming in that window and this turned the hallway into a blowtorch.

Engine Company 10 who was first due had beat us in and one of their guys named Roger had stretched the inch and a half to about the second and a half floor landing and he since he didn't have an air pack on, that was as far as he could get so he was hosing down the blowtorch from the stairwell. When we got into the hallway there was not much smoke and we could feel the heat, the heat was unusually high, especially for being a half a floor below the fire floor. I had never felt that much heat at that location. When I got up there with my line, we were the 2nd hose line in and as soon as we got water, I tried to pass Roger on the steps since he didn't have an air pack. Mike and Bryan were shoving hose to me from behind. In the shuffle to get the line up to the third floor, somehow I lost one of my gloves. Between losing my glove, and the blowtorch conditions in the hallway, I was trying to maneuver the hose with my coat sleeve pulled over my hand. I felt like I was working the line like a gimpy member. About this time some ice came through the hose line and clogged the nozzle and the water went from a good hard stream to a weak garden hose stream. Simultaneously, I ended up getting burned pretty good. When I got burned, my reaction caused me to pull my hand from my sleeve, which resulted in my hand getting burned too.

Needless to say, I had to back down the steps pretty quickly. It worked out ok, since Roger was still there with 10's line and I set my line down and he passed his line to me to try to knock down this fire from the top of the steps. We still didn't know that the wind was blowing in through the window and feeding the fire toward us. The hose stream was only able to reach the tongues of the flames; we were not able to get the water to the seat of the fire to cool off the actual stuff that was burning. The hose stream itself was fanning the fire. In essence, it was a standoff for a while.

As all fires go, eventually, it burned up enough of the fuel that we could get into the hallway and make the front room. When I got the nozzle to the doorway of the front room, the fire was over. This was one of Mike Edmond's first fires and probably not a good fire for a first fire because it made it looked like all fires were hard to put out. This is not the case. Usually we make it to the doorway pretty quickly and it's a done deal.

We had become a pretty young shift and spent a good amount of time off duty enjoying the bars, nightlife and the street scene. I did that as best as I could while still working part-time and spending time with my family. Somebody took a picture of Mike, Bryan and me one night when we were out socially after a union meeting and the picture somehow made it to the bulletin board at the firehouse. At that time, there was a movie out called "The Three Amigos" a comedy movie, I believe, and somebody had written on the bottom of the picture "the three homigos." Somehow that name stuck with us even though Gene-O and Grayson were still a part of that shift. We became a very tight-knit work group. I think we could have gone out and committed a major crime and nobody would have confessed to it. We were as thick as thieves. It made for some interesting times for some of our supervisors, because if they couldn't get along with us, we could make life very difficult for them, that was how tight we were.

Working with the same guys every day and doing stuff together off duty and getting to know each other's families was just like Sully said, it was the one time in my career that I really had to cherish. By 1986 we were in that zone; we were going to fires, having fun racing the other companies to the scene of fires, making rescues and had become a cohesive group. We all gave that extra effort to get to the scene ahead of #10 and/or #6. It was good times. I was acting (taking the role of the officer because the officer was not there) and Mikey (the one with hair like Mike Ditka) was still the new guy and it was August and the Lt. was off so I assigned the nozzle to Mikey. I wanted to see what the new kid could do and I knew this was the perfect opportunity for me to teach him a thing or two. Mikey and I were about the same age, but he had never been in the fire department before and sometimes he was a little skittish or nervous and I thought this was a great chance to show him some things. Back then you didn't take the fire engine to the store; one guy went in his personal automobile to get the groceries for our meals. On this particular day, we took the wagon and the engine to the Safeway grocery store at Harrison and Broad and were in the checkout line when they put out a house fire in the 400 block of Goshen Street for 5, 10, 41, and 52. 3 Truck was out of service because that was normally their run. Being at the store didn't put us that far out of position because 10 still had to come by us at Harrison Street to get to Goshen Street. We pulled into the block and had heavy smoke from the 2nd floor of a 2-story brick row house. Bryan was driving the wagon that day and we stopped it at the hydrant on the corner and laid out; the engine stopped put into the hydrant to pump the hoseline.

When we got there, there were people on the street pointing us to the doorway so I ran up into the house. This type of row house had a hallway that ran to the back of the house. The first room on the right was the living room the next doorway on the right would be a staircase that ran on a 90 degree angle straight up.

These houses are so narrow that when you get to the top of the steps, the stairway splits and there is a bedroom that is up two steps from the landing on the front and back. On the first floor to the rear of the house was the kitchen and bathroom. I went up the steps and the back bedroom was fully involved. I made it to the landing and saw that there were no victims on the floor. We didn't go looking for medals, but over the years, I have developed some tricks most of which I learned from riding with firemen in New York City and getting to know some of the guys.

One of the tricks was before you put water on the fire, look down, even during heavy smoke conditions, the fire will be drawing oxygen at the floor level and you will have between 6" and 2' of clean air at the floor before you put water on the fire. Victims will usually be on the floor unconscious. I checked the front bedroom and didn't see anyone either, so I scooted back down the steps and met Mikey in the downstairs hallway with the hose and guided him upstairs with the hose to the fire. A room fire is the most rewarding fire you can fight. You take the nozzle to the doorway, whip it around for about 30 seconds, and 90% of the time it's out and all that is left is to crawl in there and put out the hot spots such as mattresses and chairs that may be flaming up. Mikey did a good job. I am not sure if that was his first fire, but it was one of his first fires. We were kind of pumped up and the teamwork was starting to gel, the three homigos concept I call it.

It was a super hot day, and I remember a picture in the newspaper of Donny Viar taking a drink of water from the fire hose because it was about 98 degrees that day. Richmond summers are hot and the humidity is unbelievable. Thank God we were still wearing out ¾ boots, I don't know how these young guys today do it. We overhauled the scene, packed our hose up and returned to the store to get our groceries and then headed back to quarters.

Gene-O cooked two excellent meals that day and we were getting ready to sit down to a steak dinner when the box opened up for 5, 6, 41, and 51, 107 W. Duvall Street for an apartment fire. Duvall was a street directly behind the firehouse and 107 was about 3 blocks to the east. As we came out of the kitchen and headed for the apparatus floor we could smell the wood, plaster and lathe burning. A building fire has a unique smell all to its own which is a sign you are going to work. There is no other smell like it. To smell that before you get on the fire truck, it is always an adrenaline rush and pumps you up. We laid out from the corner of Jackson and Duvall and at the 100 block of Duvall was a 2-story cinderblock section 8 type housing building where everything in the building was masonry and indestructible and the only things that would burn in that building would be the contents. You would be surprised with only a sofa and furniture how hot the fire can be. We were met with heavy fire showing out of one of the apartments on the second floor and it had an aluminum screen door and the fire was so intense that when I went to open the screen door it folded in half because the whole top half of the screen door had gotten so soft and melted that it just folded. Mikey was on the pipe again and his eyes were big and bright with excitement and intense concentration. We hit the fire from the balcony/walkway; it was like a motel balcony setup. We knocked it down and started into the apartment. I was right behind Mikey coaching him on as we moved along with words of encouragement. It was very, very hot and intense and it was the second time we were in a good fire in less than 10 hours and we were pumped up. Soon I thought we had the first room under control, I couldn't tell because of the intense heat, there was still enough heat that I thought we had another room on fire in the back. So I grabbed Mikey by the arm of his coat and told him to clear all the hot debris from in front of us using the hose and we pushed into the back room to look for the fire. It turned out that there was no fire in the back room and the heat was just from the fire in the front room. Mikey told me later that when I told him I thought there might be another room on fire he was thinking, "Damn...."

Having two serious fires like that in such a short period of time and being brand new to the fire service, Mikey wasn't sure he could make it through another fire so soon. He told me that he thought if I could do it, then he could do it, and he stuck it out with me. That's the secret of being a good fireman; the willingness to stick it out that extra few minutes to ensure the job is done properly. It is known as "the moment of truth;" whether or not you have the courage to keep going forward when others are bailing out all around you. I think he learned some things that day.

Bryan Lam was another great fireman who came to the job with-out any firefighting experience. His father Joe was a long time fireman who was now at 16 Engine in North Side. Bryan was an electric lineman before he came to work with us and was always a valuable resource at the many emergencies we went to involving electricity. Bryan, Mike and I all enjoyed a good laugh and the type of humor found in Monty Python Movies. We particularly enjoyed one movie where the characters were galloping thru the woods on foot pretending to be on horseback, the last guy in the pack had two coconut shells and was making horse hoof print noises. It became the prop for many jokes around the firehouse.

We were sent to 35th & Broad one evening, first due on the second alarm for the house. When we arrived a large two story Victorian style house was well involved in flames. We laid a supply line from blocks away and were ordered to take a two and a half inch line to the second floor. Bryan, Mike, and I manhandled this line up the large Victorian staircase. Reaching the second floor, 2 Truck was there pulling the plaster ceilings for us and exposing tons of fire in the attic.

As soon as we started flowing water onto the fire The Battalion Chief ordered everybody out of the building. We manhandled the big line back down the stairs and outside and flowed water from the street. After a short period of time the Chief orders everyone back into the building. On this second trip we were left with-out a Truck Co. and had to pull our own ceilings. Now this job was starting to stink as the amount of physical exertion wasn't any match to the positive effect we were having on this fire.

Again as soon as we start extinguishing fire, the Chief orders the building evacuated. This time were a little bit pissed off, and to lighten the mood as we came out the front door we were all skipping as if on horseback with Bryan bringing up the rear doing the galloping sound effects and Mike and I yelling "RUN AWAY!". The Battalion Chief watches this show and comes over to our Lieutenant and tells him he has half a mind to have all of us drug tested. From that night on the evacuation signal was referred to as the run-away signal.

The fires and camaraderie made it painful to take paid vacation days off. We were having so much fun at work that you knew you would miss something if you took a day off. There are very few jobs in this world where guys don't want to take paid time off, but that's how it was during this time period.

Gene-O was our NASCAR style driver on our shift. Actually he drove fast and would scare the hell out of you, but he couldn't back up 50' without hitting something. We came into work one day and Captain Marsh announced that we were getting a new wagon. What, how could this be? Our Mack wagon was only 10 years old and it had a short wheelbase and you could put it in alleys on the first try. It was a quintessential city fire truck, no bells or whistles, but it performed like a Timex watch (It took a licking and kept on ticking). We were all very comfortable with that fire truck and didn't want a new pumper. We protested but the Captain explained that Chief Lewis had already decided, and it was a done deal because our beloved Mack had almost 100,000 miles on it. Gene-O protested that most of the mileage was from pumping water and we knew that was true because when you were pumping the speedometer averaged about 40 M.P.H. But the Fire Chief had made a decision and that was that.

Then we learned that our new wagon was manufactured by Pierre Thibault Corporation in Canada. Now we were livid as 12 had gotten a Thibault pumper the year before and the thing was an abortion. It had a long wheelbase and seemed to be ten feet wide. It would have made a decent rig for a suburban fire department but it wasn't a ghetto fire truck. So we weren't happy campers when a few days later we were instructed to meet the shop mechanic and a representative from Thibault to be qualified on the new rig. Captain Marsh was off and Bryan was in charge of the company when we went to Byrd Park to do our driving part of the training. Byrd Park is a large park with one way asphalt streets thru the park, as were driving over there Gene-O is telling us he wants to be the last one to drive the new pumper because he said he was going to try and flip it over. So now we are all smiles as we know Gene-O is crazy enough to do it. The poor shop mechanic and the delivery rep, a short French Canadian guy, have no idea what is about to happen.

We get a short course on trouble shooting the new rig, and Bryan is the first one on our shift to take it for a spin. His first lap around the park is calm as he is getting the feel of the new wagon. On his second lap he puts his foot to the floor and we can hear it sliding around the corners as the smoke from the rubber drifts thru the park. Mikey goes next and actually gets her up on three wheels turning one corner. Grayson Finner goes next and when he gets back, the delivery rep. from Thibault looks like he is about to pass out. It's my turn and I put her through the paces while the Frenchman has hold of the dashboard with a white-knuckle grip. He is begging me to slow down when I slam the brakes on and slide the rig to a panic stop. It takes twice as far as the Mack to stop and the Frenchman looks at me with tears in his eyes and asks me why we don't like his fire truck. "Because it feels like I am driving a Tug Boat," I reply, and the name stuck. From that day on we called that rig the tug boat!

When we get back to the group, the Frenchman has had enough and will not let Gene-O drive. We all look at each other and burst out laughing as Frenchy has no idea what he just prevented. Frenchy exclaims that he has delivered hundreds of new fire engines and has never seen a group of men not happy to be getting a new engine, and he doesn't understand what is wrong with us? Grayson in his straightforward manner says "Well good for you, take this piece of shit back to Canada with you then!"

There was an elderly woman who lived a couple of blocks from the firehouse and was a retired English teacher. She would bake a cake and cut a couple of slices out for herself and bring the rest of the cake to the firehouse.

While visiting with us she would constantly correct our grammar. While this may sound maddening we actually enjoyed her visits. I can't remember her name but behind her back we called her the crazy lady. She seemed just a little bit off and no-one ever ate the partial cakes she brought us. We were all afraid she was looney enough to put something in the cake. The crazy lady would visit about once a month, and after she left we would correct each others grammar and miss pronounce words like faux pas which we would pronounce Fucks-pass.

One summer afternoon an elderly black gentleman came to our door and explained the he was the crazy lady's ex-husband and hadn't heard from her in a couple of weeks and wanted us to put a ladder to her second floor window and go into her house to see if she was alright. So we take the wagon up to her house and its 98 degrees outside, I climb the ladder and open a second story window and the distinct smell of rotten flesh hits hard.

The crazy lady has passed away in her bed and her head had swelled up to beach ball size. I go into the house and down the stairs and call the police from the phone in her kitchen. The dispatcher is confused and I have to repeat my request a couple of times as she keeps telling me that 5 engine isn't on a call.

When the police get there we take down our ladder and tell the elderly gentleman we are sorry for his loss, but we too feel the loss as although she was unique, we too will miss the crazy lady…..Oh and pardon my fucks pass!

- Chapter 16 -

Jackson Ward

The neighborhood I was working in was known as Jackson Ward, named after President Andrew Jackson It was also known as the Harlem of the South, due to all of the prominent black people who lived and visited the neighborhood.

Richmond is full of history; the main street through Jackson Ward was Second Street. It was known among the black community as "Two Street." A lot of prominent and very successful business people had establishments on Two Street. Many of them survived into the 1980s when I worked there. The Eggleston Hotel at Second and Leigh had pictures on the wall of many prominent black politicians, jazz artists, and clergymen, Baseball players from the 1920's, thirties, forties, and fifties. But, with a good bit of crime in the eighties, Jackson Ward was in decline and crack cocaine had taken over the streets along with other drugs and illegal activity. It was a rough and tumble neighborhood, but as I worked there, I really started to feel at home.

It's amazing, as a fireman you really have two lives. You have your life off duty with your family; I lived in a town about 7 miles North of Richmond named Mechanicsville with my wife and young son in a predominately white neighborhood, in a brand new two bedroom, one bathroom house with an unfinished upstairs where I would later put another two bedrooms and a bath. I got to know my neighbors in Mechanicsville, and that was very interesting, but I really got to know my neighbors in Jackson Ward much better than I knew my Mechanicsville neighbors. This was simply due to the fact that in answering calls for help you would actually go into their houses and see how folks lived.

The fire house was located at the corner of Brook and Leigh Sts. and had houses and apartments all around it. The children would play on the streets, and it was interesting, during my eight years at that house, I really watched those kids grow up in front of my eyes. You got to know them, and you could pick out the mischievous ones, the ones you knew were headed for trouble; but also a lot of good kids, who tried hard in school. There were a lot of single parents, women who were trying to raise their kids the best that they could on their own. There were also a lot of knuckleheads, drug addicts, thieves. I would like to tell you about a couple of them who really made life interesting at work, in between all the activities, as far as emergency calls, fire alarms and what not.

A call came in one day at Second and Leigh Streets for an automobile fire. We responded, I was the acting Lieutenant that day, and we arrived at the corner to find an older model Buick, with the interior filling up with smoke. There was an elderly black woman sitting on the bench at the bus stop, just observing things and watching people walking up and down the street.

It was a sunny afternoon; nothing appeared unusual, just another day on Two Street. Anyway, as we pulled up, it was apparent the car was parked by the curb, with nobody around it. We tried the door handles, they were locked and as I said the car was filling up with smoke. So one of the guys went over to the pumper, got a pike axe and proceeded to break the driver side window. Now, back to the elderly black woman on the bench, she appeared to be somebody's grandmother. She looked like a distinguished woman, in nice clothes, just sitting on the bench and taking it all in. Now when Gene-O swung the axe, he swung back like a baseball bat and proceeded to break this window in full swing, which alarmed the lady that we will call "Mom Mary". Mom Mary let out a vindictive swear that would make a veteran sailor blush. She started screaming, "Did you see what that white mother fucker did?" "Did you see what that white mother fucker did?" on and on and on. Well, we had our hose line stretched and were proceeding to try and put this automobile fire out, and really weren't paying much attention to Mom Mary. It was almost comical because as I said to look at her, she could be someone's grandmother there, and then when she spoke it was the last thing in the world you would expect to come out of her mouth. But, looking back, the fire crew was all white in a predominately black neighborhood and whether she knew the owner of the car or not, I don't know. But I don't think in her mind she thought we were treating that property with the respect and dignity that it deserved, even though it was burning up as she sat there watching it.

As we got into the car, the fire was behind the dashboard and in the engine area, and we opened it up and extinguished it with no problem. But, it was Mom Mary that wound herself up with just continually calling out obscenities, such as "Did you see what that white mother fucker did?" and "These white mother fuckers don't care about us" and she was starting to draw a crowd. The fact that people were stopping on the corner to see what this woman had to say, you could see it in their faces that they were looking for something more than what they saw, as far as the fire department simply putting out an automobile fire.

In short, it was the beginning of what could potentially become a civil disturbance and it was totally unnecessary. So, having five people on my engine and being in charge that day, I walked over to Mom Mary and engaged her in conversation. I tried to explain to her what we were doing and why it was necessary to break the window of the car. Even though Gene-O was a little dramatic with the way he swung the axe, it was necessary. She started to calm down, so I got her to sit back down on the bench and we were chatting and this woman was probably in her sixties. Finally, she looked at me and after talking to her for a few minutes it was apparent that she might be a little bit intoxicated. She liked her wine. Soon the crowd began to dissipate, the car fire was out and I was trying to engage in what I call community relations. Finally, this woman looked at me and told me her name was Mary; she looked at me with a straight face, right in the eyes and said, "You know, I've got a son that looks just like you". Hence, the name, Mom Mary. It began a relationship that lasted the next four to five years. Pretty much every time we passed Second and Leigh, this was still back in the days when we hung on the back step of the wagon and were in exposed positions, every time I saw her I would shout out, "Hey Mom Mary!!" and she would shout back, "Hey mother fucker!!". If by chance we happened to catch the red light and had to stop at the intersection, she would always make her way over to the wagon and bum some money for some wine. I don't know how many five dollar bills or one and two dollars that I gave to her, but I always felt a special kinship to Mom Mary. It was sad, we probably saw her for the next four or five years, and eventually we didn't see Mom Mary anymore. I imagine that she'd passed away or got too old and infirm to make it back out on the street. I never really learned her real name, but it was an interesting relationship, and as I said a lesson in community relations.

Another character in Jackson Ward who used to happen by our fire house almost every night about the same time, between eight and nine, was a black gentleman by the name of Shirley Johnson. The reason I can tell you I know Shirley Johnson's name is because he would come by heading east on Leigh street, completely intoxicated, talking, singing, and carrying on. He was always well dressed, in a jacket and tie, albeit a little outdated, about twenty years out of style. But Shirley would always stop and sing a song for us followed by "My name is Shirley Johnson", as he pulled out his identification card, not a driver's license, but a State of Virginia ID card that was issued for him. I suppose being in uniform and sitting in front of the firehouse we looked like we were cops or something, but he always felt obliged to identify himself and show his ID. It was comical in a way because after a period of time everybody knew Shirley, we'd see him coming two blocks away and we would always have something to ask him or someway to tease him about things that we knew were going on in his life. He was part of the entertainment of the neighborhood, it was almost like a daily ritual, and he was always right on time.

A third character that used to come into the firehouse frequently was a guy we called Shorty. I never knew Shorty's real name, but he was a short man, about 5'2", had snow white hair, coke bottle glasses and was sort of pudgy. Now Shorty's claim to fame really happened before I came there, back in the day when they still had street corner fire alarm boxes on each corner of the City of Richmond. The story goes that someone pulled the box at Price and Jackson and 5 responded. It was snowing like crazy that night, a cold January night, when the wagon arrived at the corner at Price and Jackson the snow was completely undisturbed expect for the set of footprints that came out of the front door of a house on Jackson Street, right to the corner box, and back to the house on Jackson Street. Further investigation revealed that that is where Shorty lived but Shorty denied everything, he didn't pull the box, but the footprints matched his shoes and so on and so forth. Anyway, he was prosecuted for pulling a false alarm, but for some reason over the years the guys forgave him and Shorty became a frequent visitor, always entertaining and engaging. In fact he used to drag his lawn mower the five blocks to the firehouse to mow the twenty by twelve foot area of grass each week.

As I mentioned before, Jackson Ward was predominately black, it was poor, and there was an element of crime in the neighborhood and the Gilpin Court Housing projects were only two blocks behind us. You had to admire and respect the women trying to raise children in the apartments directly behind the fire house. Another thing, looking back that amazed me was how we watched the children grow up playing in the alleys behind the firehouse. We did our best to be good neighbors, fix broken tires on bicycles, fill their tires with air, and so on. But some of the kids as they got a little older became a pain in the butt. You could pretty much tell the ones who were going to get into a life of crime, and unfortunately, we ran some of them on shooting calls later on when we started Emergency Medical Services.

I was always amazed as I got to know my neighbors in Mechanicsville at their disbelief. They couldn't understand how I could work in Jackson Ward, it's so dangerous. But really, the longer I stayed there, the danger faded and the dirt and the crime didn't seem so bad either.

Amazingly, in twenty plus years of service with the City of Richmond, I never had my vehicle broken into, even though many times I came to work in the morning and there would be a car sitting on cinder blocks or a window broken out where someone had gone in and stolen a stereo or something. Sometimes all that would be left was a pile of broken glass on the sidewalk and the car itself would be gone, stolen for an overnight joy ride. Twenty-two years, I never had that problem. I never left anything in view that someone who's walking through the alley, cold and hungry and looking for a reason to steal something would have a reason to break my window.

It was about 1985 when we ran a call in the 700 block of North First Street when I saw empty vials on the sidewalk. During this time the crime was really starting to take off, you could tell that the neighborhood was changing, and that the young kids didn't care. People were getting shot over stupid little things, and I will never forget we were on the scene of an overheated oil furnace fire and we saw these little empty plastic vials on the sidewalk and I just spoke up and I asked what they were. A policeman within earshot just looked at me and laughed and said, "That's crack cocaine kid. That's what they keep their crack rocks in. Why do you think these people are going crazy?"

It was late December and there was an elderly gentleman whose house we responded to he spoke up and, most of the people on this block owned their own homes actually I think that it was one of the only blocks in Jackson Ward I never responded to a working fire in. But this gentleman spoke up and said that all their lives they had worked hard, struggled to purchase their own home and take care of their families, and most of them in that block were successful; but now seventy years old, he had to barricade himself inside his house every night, he was scared as hell. Gunshots erupted; people were doing drugs, and dealing drugs directly in front of his house. NO matter how many times he called the cops, nothing changed, nothing ever went away, and he directed his anger at the policeman and it almost started a fight. I looked back and I smile because the fight almost erupted because I had just inquired was what were these empty vials on the street.

Now as you read this book, you probably think that life was easy and that everything was as simple as pie and I guess with time you forget the hardships, but as we are talking about life in Jackson Ward and Mechanicsville; a fireman, at least back then, didn't make much money. Of course, the work was rewarding, we were helping people almost every day, but the money really didn't go very far, especially with a baby at home and another one soon to be on the way. However, working 24 hour shifts ten days a month afforded the ability to have a second job, a part time job. My first part time job I got right after I got off of probation. It was driving an oil tanker for a home heating company called Woodfin Oil Company in Richmond. Woodfin was a pretty large oil dealer and had other several firemen working for them part time. Dale McMillan or Mac as we called him was a fireman at 17 Engine, in South Richmond, and had been with Woodfin and its predecessor companies for at least 20 years. He broke me in driving the truck, it was a chance to make some extra money on the side, and these trucks also had the same wheelbase as the pumper and a higher center of gravity, so by learning to drive the oil tanker my ability to drive fire engines was enhanced.

The oil tanker as I said was heavier, and also had a higher center of gravity. After two months of driving the oil tanker, driving a fire engine was actually pretty easy.

The other plus to this part time job was I got to learn the city streets and alleys intimately. There wasn't a part of Richmond that you could take me to that I didn't know where I was at. In fact, in some cases I knew the alleys better than the streets.

As I said, Mac worked there and broke me in, but there was also another fireman named Woody Wilhelm who was a lieutenant out in Henrico County. We enjoyed working together and had a pretty good time. After about 3 winters with Woodfin, I had the opportunity to go drive the big gasoline tanker trucks for a company named Eastern Motor Transport. In fact, Woody went over there first, and invited me to join him. These were the kind of trucks that delivered to gasoline stations, big heating oil companies; stuff like that, the 9,000 gallon tanks.

There was a fireman who worked for D.C. who worked for Eastern, I heard about him for years and years, a guy named Timmy Jones, and Timmy and I later became good friends. But it was years before I ever met Timmy, all I would hear about was this fireman from D.C. that was a good worker and crazy just like me. These gasoline tanker truck drivers were known as suicide jockeys and several of my friends used to kid me that between 5 Engine and Eastern, I was trying to kill myself.

The other part time job I worked during that time was with Gary Baker. Gary was a carpenter by trade who also happened to be on the same shift at 11 Engine Company. I got hooked up with Gary while trying to find somebody to help me finish-off the two bedrooms and a bath on the second floor of my new house in Mechanicsville. While I have never been much of a carpenter, I could tote lumber and be the extra set of hands needed to lift heavy objects. He had a small business on the side framing houses. It was good honest work and good exercise. Working with Gary taught me many things about working with my hands and would be invaluable at fires when the subject of building construction arose.

That was life for me in the 1980's, pretty much consisted of working 24 hours at the firehouse, getting up the next morning, building houses or, driving a truck all day, getting home late that night to a nice hot meal cooked by my wife, some quality time playing with the kids, and off to bed and back to the firehouse the next morning.

It was during this time I started to think about climbing the promotional ladder. I had worked for some great officers, and I had endured some clueless cowards too. Maybe it was time to try a command of my own. It was sort of a crazy, hectic time but life was very good indeed, again everyone should be so lucky.

- Chapter 17 -

Guardian Angel

Most people who know me would probably be surprised to know that I have pretty strong religious beliefs; I don't wear my religion on my sleeve, but throughout my career it seems like we were constantly getting into situations that could have turned out very, very badly if it hadn't been for what I believe was a guardian angel looking out for us.

This particular fire illustrates exactly what I am talking about. It was a daytime fire, it came in at the corner of Lombardy and Grace Streets as a building fire with people trapped. We were running 3rd due engine and when we arrived at Lombardy and Grace both 10 Engine and 3 Truck had all hands throwing ground ladders to the 3 story brick row building with heavy black smoke coming from the first floor through the roof. It appeared that the fire was somewhere on the first floor.

We ended up laying a line from 10's engine which was put in at the hydrant at the alley between Broad and Grace on Lombardy Street. We laid up the street and since the front of the building was so congested with 3 or 4 ground ladders being raised, multiple people being led down ladders through the heavy smoke conditions it was more advantageous for us to lay past Grace Street and up the alley to the rear of the building.

This building was about three buildings from the corner and when we got to the rear of the building with our inch-and-a-half we found the rear, 1st floor room fully involved. I kicked the door in and in a minute or two we had water and I had the nozzle and knocked the fire down from the alley and proceeded to crawl up into the room. Grayson and I worked in there and as we advanced toward the front of the building, it became apparent that we had just a one-room fire. We began to throw the mattresses and over-stuffed chairs out the windows and when the smoke cleared, this became one of the strangest fires I have ever been to.

Most fires that start in a building usually burn up and out, but this particular fire, once the smoke cleared and we looked up, about 40% of the floor above us was gone and it appeared after further examination, that this fire began as a mattress fire in a room on the 2nd floor that had burned through the floor and dropped down into the 1st floor. There was very little fire damage to the 2nd floor room other than heat/smoke. This fire actually burned downward.

What made this interesting and made me believe that the good Lord was looking out for us was that there was an old cast iron radiator on the 2nd floor that was being held up by only by 1/3 of a 2 x 12 joist at one end and the supply pipe to the radiator and at the other end. This radiator was located directly above the door where we entered the building. If that had collapsed while we were underneath it in the smoke filled room, it would have crushed us. That was the one of many examples of my belief of divine intervention in action.

Many times we would get into a situation at a fire, maybe being above the fire, or being cut off by fire, or just in a situation that I knew we were in trouble and I would look at whoever I was with and say, "That's it, we're screwed now…" When you are in that situation, you are thinking that you are going to die right there. The job is that intense sometimes.

I mentioned that one of the best officers in the fire bureau was Lieutenant Dave Pulliam. I went to many fires with Dave and even with the dirt and the danger I can't underscore how dangerous it really is. Dave was forever getting out of the front seat of the wagon and running into the building with only his helmet and coat on. For some reason, he never liked to wear his boots (and air-packs were for sissies). He would have on his uniform pants and patent leather shoes.

Somehow Dave was always able to get up there and make rescues where someone who had full turn-out gear on would have had difficulty performing the same feat. Dave moved like a cat, he was fast. I have seen a number of times where the difference between life and death in a fire is just seconds. I can't stress that enough. Sometimes firefighters today forget that when a human being is in a burning building, they don't have a lot of time. The fire is expanding exponentially and from the time somebody calls 911, to the time when the call is dispatched to the time when the fire department arrives and gets water and a line into the building, there is not much time for the people in the building. That's one thing that made it interesting working with Dave. He broke all the conventional rules as far as safety, but always seemed to pull it off somehow.

I remember one night working with Dave and we went to Bowe Street, the 800 block, for a building fire with children trapped. I was riding on the wagon and as we were coming down the street that night, each man was trying to get a game plan in his head as to his role when we arrived at the scene. Engine 10, 43 and the 2nd Battalion Chief were also on the assignment. We arrived at about the same time as #10 and when I came off the back step with the inch and a half, Dave yelled for me to bring the line and he was going to find the kids. Now this house at the 800 block of Bowe had heavy flames blowing out of every window on the 2nd floor; it was lit up like a roman candle.

Dave in his normal MO takes off through the front door in his coat and helmet. By the time I got halfway up the stairs with the line here came Dave on his way back down the staircase with a young black child about 5-6 years old in his arms. The child's clothes were half burned off of him and he was not breathing, but Dave had begun mouth-to-mouth on his way down the steps. I had to hug the wall to let him get by and when I got to the top of the steps, the second floor was nothing but a sea of flames.

It was almost like a Hollywood movie. In most building fires you can't see two feet in front of you, but here it was just a sea of orange and you could see everything in front of you burning. Once we got water, Bryan and I proceeded to start knocking down the fire on the 2nd floor. It was one of those fun fires, where we got to go from room to room knock the room down, back out, and proceed down the hall to the next room and knock down the fire, and so on. By the time we got to the third room our strength was sapped, our bodies exhausted. I remember trying to get to the fourth room; my mind was willing but my body didn't want to go. Bryan moved up and took the line from me, and I leaned up against the wall like a wet, limp dishrag and I looked down at the hose line and bent down to try to get Bryan some more line.

About this time the guys from 10 and 3 Truck had arrived at the top of the steps and it got really tight at the top of the staircase and we couldn't do too much because it was too crowded up there.

After we got down from the fire, Bryan and I went looking for Dave and they had loaded the young child in the ambulance and taken him to the hospital. Amazingly enough, this child lived and Lieutenant Pulliam won an award for this rescue. I watched Dave pull off some things during my career that were simply incredible.

Jake Rixner

Back to Catherine Street, remember Catherine Street? The street where everyone appeared to be drunk? In the 1100 block there was a house that we seemed to be called to about every two months or so. We were there so often that we got to know the couple that rented out the first floor of this two story house. The older gentleman in his late 60's seemed to be the only sober person on Catherine Street at any given time and his wife was a full-blown alcoholic. She weighed about 350 pounds and was about 5 foot 8 inches; she was almost as round as she was tall. One of the reasons we went to this apartment so many times was that they had no central heat. They had a front room, bedroom and a small bathroom. In the bedroom there was a king size bed that took up 90% of the floor space in the room. Since they didn't have any heat, they had a 55 gallon oil drum sitting directly on the wooden living room floor with a jury-rigged flue pipe that led into the chimney of a fireplace that was very unsafe looking. They were trying to get by the best they could and that was the only way they could heat up the apartment.

There were 2 or 3 slumlords in this neighborhood and we knew their names well because it seemed that every fire we went to, one of these guys owned the property and we had their information back at the station so we didn't have to look it up after these frequent runs. So, back to this particular run. We were called there for some reason (I don't remember exactly what it was) and somehow Dave ended up between the heavyset woman (who was in full party mode) and the bedroom door. The woman, being drunk seeing Dave with his perpetual smile and ended up taking a fancy to him. She used her big belly to belly flop him, she hit him with her big belly and knocked him backwards onto this huge king size bed. At this point she leapt on top of Dave and began grinding on him with her pelvis. Dave was laughing and there were 2 engine and a truck company all in the living room and pretty much everybody could see what was going on. She kept telling Dave, "I'm gonna give you some of this, I'm gonna give you some of this..." Dave is still pinned underneath her and you can barely see him at this point. He reached his head around and begged the guys to help him and get her off of him. It was so comical that none of us could help him because we were laughing so hard! Once he saw that we were not coming to help him, he got really frustrated which made the whole incident that much funnier.

We went to tons of fires together, but another fire that I remember working with Dave that comes to mind was a fire at 1100 block of Clay Street. It was during the winter and each shift was catching at least one fire, sometimes two or three in a 24-hour period. It got so brutal that the air packs were frozen and wouldn't work right. Anyhow, we pulled up to the 1102 W. Clay Street one night and it was a 2 story house with an English basement. An English basement is a basement that is partially above ground. This basement was about 6' above ground and 2' below ground. The building appeared to be a 3-story building but in actuality was a 2-story building with a basement. There was a one story staircase outside the building that you had to climb to get to the front door.

When we pulled up, it was about 2 or 3:00 in the morning, and we had heavy fire coming from all 3 levels. This was also a row house that was connected on both sides to other residences. Dave and I came off the wagon, Dave in his normal helmet and coat (it was about 20 degrees outside) and as soon as I got water, I tried to knock the fire back from the front door from the street since we couldn't get to the door because of the flames. One of the basement windows was directly below the front door and the fire in the basement was coming out of that window preventing us from entering the house through the front door. I put the nozzle into that window hoping to knock down the fire enough that we could gain entry. Each time I knocked it down and went back to the front steps to go to the door, the fire would come out again and threatened to cut us off.

It was a standoff for a few minutes until #6 got there and pulled a line in the alley and knocked the basement down from the rear. What we didn't know at the time was that they had direct access to the building from grade level from the rear. Once 6 got some water on the fire, we were able to get through the front door. By that time my air pack seized and I couldn't get any air so I looked at Dave who didn't even have an air pack on, and knew that Dave loved the nozzle and if I gave it to him, he would not give it back, so I was hesitant to give it up. Instead, I ripped off my face piece and since the fire was blowing out every window, there was a pretty good air supply at the floor in this place. It got a little tight with 6 below us and the steam that was generated when they put water on the fire in the basement, and that would take our breath away from time-to-time, but Dave and I pushed this inch-and-a-half through two stories of this house.

Remember, four rooms is usually a firefighter's limit when it comes to room fires, due to the intense physical exertion, heat and stress on his body, so at 3 a.m. it really taxes your body. You go in seconds from a resting heart rate (asleep in bed) to a heart rate comparable to running a marathon coupled with 80 pounds of gear on that weighs considerably more once it is wet.

So Dave and I pushed the line through this house and after we put the 1st floor out, Radar from #6 came up the stairs and his air pack had broken as well. I asked if they had any more air packs that still worked and he shook his head no. We started up the stairs to the 2nd floor and about halfway up the stairs the breathable air at the floor went bad on us. I had my face on the steps and was hacking and coughing. Dave was lying on the stairs next to me doing the same thing. Finally, I found a pocket of air, got a good breath and advanced three or four steps quickly to get more of the fire knocked down so we could find a window so we could breathe. I looked back at Radar and saw him shake his head as if to say he couldn't make it, he wasn't coming with us, but like the great guy he was, stayed there and shove hose up to us. We made the 2nd floor and after a pretty good beating, we finally got this fire knocked down. During this whole interior campaign, the guys from Ladder 3 had put a ladder to the front porch and they were up trying to get into the 2nd floor to make a search. By the time they got to the 2nd floor, the window glass failed and they were cut off by a wall of flames and couldn't get back to their ladder. Calvin Coleman from 3 Truck was burned pretty severely on the backs of his hands through his gloves. This fire melted parts of helmets and stuff like that before Sam from 3 Truck could get another ladder up to get them down. It was a pretty exciting fire all the way around. A lot of weird stuff happened. Once again, our guardian angel was definitely riding with us that night.

Too many things happened in my career to believe that it was just luck or chance. Another example of a rescue that Finney and I made and to this day one I can't explain, it was one of those gut instinct moments was when we got called to the 500 block of Monroe Street (a daytime fire) for a house fire; 5, 6, 41, and 51. Bryan was pretty new to the company and was driving the engine that day and I was driving the wagon and Finney was on the back step. I don't remember who the officer was. From #5's firehouse you could throw a softball to 500 Monroe Street; that's how close it was.

The house was showing light smoke conditions; it looked like a food-on-the-stove fire. We had no information from onlookers, no information from dispatch, but for some reason I still cannot explain, I went up and tried the front door. I found the door was locked so I kicked it in and immediately ran up the stairs toward the origination of the smoke. (This time it was I who had only a coat and helmet on.) At the top of the stairs were three bedrooms each of which had a padlock hasp on the door which indicated that it was a low-rent rooming house. If you were home, you padlocked the inside of the door; if you left, you padlocked the outside of the door. When people were home, you could still open the door about an inch and see inside the room. The first door I came to was locked from the inside, but was ajar. I could see inside and the mattress was on fire. Usually a mattress fire is not a big deal, but what made this a big deal was what was inside was a man, about 80 years old who had gotten his torso off of the burning mattress, but from his waist down, his clothing was on fire. Finney was right behind me with the hose line, but I didn't wait for the hose line, I kicked the door in, breaking the padlock from the door, grabbed the guy under his arms, and dragged him out of this fireball and Finney was right there to help me put the fire out on this man with our gloved hands. The man was moaning and was in shock and in a lot of pain. I picked him up under his arms, and Finney got under his knees and we carried him down the steps to the amazement of Lieutenant Bridges who was from #6 and who we almost ran into at the bottom of the stairs.

The victim was burned with 3rd degree burns from his waist down and with his age it may have been a fatal injury but we had to give him our best shot. We called for an ambulance and this was one time that I stayed with the patient and took water from the booster line and tried to cool his legs off. A severe burn like that is like a pot roast in the oven... it will continue to cook even after it is removed from the heat. I don't wear my faith on my sleeve, but too many things have happened for me to believe that it was just me calling the shots, not the man upstairs.

Another good fire we went to during what I call "the good years" which was something that Sully, a good buddy of mine from Maryland, used to say was when you worked with a good group of guys and everything just jelled and went smoothly. When you experience that, he said, enjoy the moment because once transfers or other changes occur, you probably will never see that moment again in your career. He had that kind of group when he was the Engine captain at Kentland and I could see that coming to pass. Mike, Bryan, Gene-O, Finney and I were such a well-oiled machine that we didn't need to speak to each other in order to communicate. We all knew what each other was going to do. Some people like the opera or performing arts for the sheer inherent beauty, but to see 5 or 6 guys working together that can perform like that; it was a beautiful, well-orchestrated thing. That's not to say that we didn't have our mistakes or never screwed up, but the fact was that we were going to so many fires, if we went a week or two without a good fire, you could see some of the inefficiencies or rusty areas that haven't been practiced in a while. Little things like trouble stretching lines or forcing doors would happen after there weren't any fires for a while.

I made a bunch of rescues and assisted Finney and Bryan on some rescues and so it wasn't all about me. We were summoned to a fire one night at 29th and Broad Streets and we knew that since we were called on the 2nd alarm to Church Hill, there was some quality work waiting for us since Church Hill had some of the best firemen in the city.

There was a 2 story row house with a large bay window and the end unit was on fire and the fire had extended through the attic to the 2nd house on Broad Street. We were told to pull a line into the adjoining building and pull the ceiling to try to cut the fire off, which we did. I was bringing the nozzle up the steps and Bryan and Finney were helping me, Finney was driving the engine that night and by the time I got to the 2nd floor, Mikey had found a woman who was about 5'4" about 200 pounds, who was burned and overcome by smoke in this exposure apartment from a couple of the rooms that had ignited in the rear which couldn't be seen from the street. It just looked like an attic fire from the street. What made this comical was that every time Mikey grabbed a hold of this woman, she was buck naked and her skin was very greasy—it looked like someone had coated her with Crisco oil. Any place you tried to pick her up she would slip out of your hands, like a greased pig. Watching Mikey trying to get a hold on this woman and get her across the floor was almost comical since he was silhouetted by fire as he tried to pull her toward the staircase at the front of the building to safety.

Bryan went to Mikey's aid, and I worked the line to knock the fire down.

Once we got the 2 rooms knocked down, the truck company came in and pulled the ceilings for us so we could get to the rest of the fire. I believe the woman lived which is always icing on the cake…when we can make a positive difference, after all that's what being a real fireman was all about!

- Chapter 18 -

Studying for Promotion

As I stated before our shift at 5 Engine had become as efficient as a well oiled machine, and life was great. I wanted that experience to last a lifetime, but I also knew all good things come to an end. To take the Lieutenants promotional exam you had to have five years as a Richmond Fireman, my time in grade would be up in June of 1987. An exam was announced for 1987 and the motivation was high to hit the books. There were 13 books the exam was made up from. Things like basic firefighting practices, to advanced thermal dynamics, as well as building construction comprised the material in the test, and if you stacked the books up they stood two feet high. In short there was a lot of material to know. I hit the books the way I hit anything I was serious about, wide open, and soon every minute that we weren't busy at the firehouse I was off in a corner with my nose in a book. It was ironic, I hated school and never even considered college, but now I was like a pre-med student into the books at every chance.

In my short time of service to the citizens of Richmond I had been inside of hundreds of building fires, and worked with or for about 50 different officers in the Fire Bureau. Most were outstanding Like Captain Emerson or Lt. Pulliam. But several were outright cowards or worse, clueless. I was motivated by both types of leaders; the good ones taught me how to lead, while the bad ones taught me what not to do!

During this time period a young black Lieutenant by the name of Val Jean was promoted out of 13 Engine and assigned to our company. Val

Jean was always well dressed, very refined and exuded an air of sophistication, so naturally the ladies loved him. However, he was a married man with a beautiful wife we called Red who kept him on a short leash. This made for some funny stories. Val Jean loved to laugh and play so he was an excellent fit on our shift, but he knew when to separate himself from the guys to take his position as the leader of the group.

One of the fun things of working with a group of men like this was the daily jokes that evolved into a language of its own. Like the "run away signal" we practiced at the few fires we were told to back out of. Bryan came up with new nicknames for the rest of us after Val Jean arrived. Everyone was nick named Jean. Mikey became Green Jean because he was a true country boy. I became blue Jean because of my blue eyes. Grayson became nutra Jean although he didn't have dandruff, while Gene-O became Jean squared. Bryan gave himself the nickname electra Jean because he had been an electrical lineman before coming on the job here. Other officers and other companies looked at us as if we were smoking crack when we were working at fires until they picked up on what we were doing. "Lt. Val Jean would you like Green Jean, Nutra Jean, Electra Jean and me to take this hoseline to the second floor and extinguish that fire?" "Boy that Gene squared drove awfully fast today, I almost soiled my jeans". "Are you a good Jean or a bad Jean?" And the fun and games went on everyday.

Lt. Val Jeans's partner on the same shift in 1 Truck was Captain Holt. He too liked to keep the mood light as long as the work got done. Going to as many fires as we were it was inevitable that guys would get hurt from time to time. Capt. Holt brought a little kids plastic cowboy toy into work and anyone who marked of sick or was injured on the job returned to work to find this plastic cowboy "Award" taped to his locker. You kept the award until the next guy marked off or got hurt.

One of the dirty little secrets of working in a group this tight was you start to bend the rules a little bit. We never drank beer at work as it was against the rules unless there wasn't an officer working. Bryan was the normal wagon driver when we had an acting Lt. As he would drink two beers and stop. The rest of us saw no problem having a few cold ones on a hot summer's night. With Lt. Val Jean on a vacation day, all the other Jeans were sitting in the alley behind the firehouse one hot June night enjoying some forbidden drink.

We had just gone inside to go to bed at about 0300 Hours when the box opened up sending us to a third alarm of fire at 1310 Hull Street in South Richmond. As we crossed the Manchester Bridge we came up with a plan of not asking for orders and taking the alley so as to stay out of sight. We arrived at 13th and Hull and laid a line from 13's Engine and up the alley to the rear of a large two story furniture store that had smoke billowing from everywhere. There was another Engine Co. and a Truck Co. in the alley so we had to hand stretch the last 100' or so to get to the rear of this building, between us and our objective was a couple of rear yards/ lots, separating these lots was an old snow fence. A snow fence is a wire fence with wooden slats in it, this particular fence was old and the kids had stomped it down so as to cross the fence. Grayson had the pipe and attempted to cross the opening in the fence when his ankle caught the wire in the fence and he fell down while holding the big two and a half inch line. As he attempted to extricate himself from this situation, the fence continued to wrap him up, the more he struggled the worse it got. The rest of us saw this and found it incredibly funny and were watching him struggle as we laughed. He was begging for us to help him when someone cleared his throat loudly behind us and we turned to see Battalion Chief Beaveman standing there with a look of contempt on his face. It was obvious he knew what we were doing and we all figured we were in big trouble. We pushed our line into the first floor rear and as soon as we started hitting fire the Evacuation signal was given. There was no "RUN AWAY" nonsense tonight as we knew not to push our luck. We held a position in the alley, and lobed water into the second floor windows until the sun came up, and I remember swearing never to drink beer on duty again.

The combination of a smoke headache and being hung-over with no sleep was miserable. Chief Beaveman came by our firehouse the next day we worked and asked Lt. Val Jean to speak with me. The Chief took me out back to the alley and told me that he knew exactly what we had done, and how disappointed he was. I apologized and told him it would never happen again, and I asked him why he singled me out? Because you're the group leader, the alpha male, the reason this Engine Company, on this shift has gone from being one of the worst Companies in the bureau to one of the best. Then he asked if I was studying for the Lt.'s. Test. I couldn't believe my ears, I went from being afraid of being written up with a risk of losing my job, to being encouraged to make rank. It was a moment of truth for me and I swore never to drink on duty again.

I asked Chief Beaveman if this was the last time I would hear about this and he assured me this whole thing was over. But he told me something I'll never forget. He said if I applied myself and studied, he knew I'd get promoted. He also told me firemen were notorious gossips and that I'd be tested very early after being promoted about drinking on duty. He just asked that I set a tone that I wouldn't condone it, but that the first time I ever caught a man breaking any rule that I had broken as a fireman, to pull him aside and give him a warning. To prefer charges on the first offense would be considered hypocritical by the men. The second offense, throw the book at 'em, you already gave them their chance. I thanked him and we went inside

We were all in bed a couple of tours later when at 0320 hours when the box opened up for a fire at 1627 West Main Street, The Trolley Restaurant. Only it wasn't for us, the assignment was for 12-10-6-43-41-52. It was rare for the truck Company to catch a run with-out us going too, and the dispatch sounded like a good job so I got up and went to the fire radio at the watch desk and listened. 12 went on scene with a two story type 3 (brick and joist) with fire out every window on the second floor. In short order 12's Engine driver reported a dead hydrant, 41 ran over 6's line and it got tangled up in the rear dual tires pulling the hydrant out of the ground (Radar was driving 6's engine and you could hear in his voice that he was still half-asleep as he called for 41 to stop on the radio), and 10's pumper died on them. I knew we were going to work when the Battalion Chief requested the second alarm.

I was driving the wagon and Lt. Val Jean was in Charge. I had the wagon barreling down Belvidere Street when approaching Main Street I noticed that Val Jean is just sitting in the front seat still half asleep, not getting dressed. You could see the smoke drifting under the street lights and the smell of a good fire hits us as we pass Monroe Park. Val Jeans eyes open and he looks at me and says, "Jake I smell smoke, I think we have fire!" I look at him as if he's on crack and he scratches his forehead and asks our due. "Well Lt. since we are first due on the second alarm and all the first alarm engines are having water supply issues, you better get your shit together." I then tell him where each of the 3 engines on the first alarm laid out from and how I am going to turn down Lombardy St. and run Cary St. the wrong way for one block to pick-up the closest unused and hopefully working hydrant. Lt. Val Jean agrees and asks me if I am ever off my game, I just chuckle as that what makes this job so great, even when the boss is off a little, someone will step up and fill the gap. We lay our line from Vine & Cary Streets one half a block to the rear of the fire building and fire is indeed blowing out of every opening on the second floor of this building and we wind up getting the first sustained water supply on the fire.

In typical Richmond fashion Mike, Grayson and Bryan quickly turn all of those orange flames into steam and soon we are picking up wet dirty hose. That building still stands at the corner of Main and Vine Streets as is still used as a restaurant. That is what being a city fireman all is about!

As I said I was studying very hard and now I was really starting to notice how all the officers I came into contact with conducted themselves. Because of the way I was raised by my father, I always disliked the officers who yelled and screamed, or tried to lead by intimidation. My father was a man of action, he used very few words. He told you once and usually the second time he spoke with his fists or would put a foot in your ass. I remember asking my dad once about being at a fire where there might be explosives or other unknown dangers. He looked at me with a straight face and said "that's why you need to put the fire out quickly before it has a chance to blow up." One thing about being a fireman is that you never know what you're facing during a serious fire.

There was a captain at 10 named Majors who was one of these guys who would hoot and holler and get all excited at fires. He had embarrassed me a couple of times and I decided to get even with him the next chance I got.

The fire in the 500 block of Goshen St. came in about 0400 hours. 5-10-6-41-43-52 was the assignment. I was driving the wagon and we could see the glow in the sky before we left the front ramp of the firehouse. Three two story wood frame houses were fully involved in the middle of the block.

There was a hydrant at the corner on Clay and Goshen and I hooked the wagon up to it with a soft suction hose. A second alarm was sounded and the battle was on. It was an engineman's fire as there wasn't much truck work to do and both 1 and 3 trucks took additional two and a half inch hose lines off of my pumper. After supplying everyone with water I could see fire spreading to a fourth house thru an alley that was lined up with my pump panel. I took one fifty foot section on inch and a half off the crosslay and connected it to the # 2 discharge on my pump.

Standing at the pump panel I could keep the side of the forth house wet and keep it from catching fire while tending to my other duties. Captain Majors came running up to me screaming for me to give him the line as he was going between the 3rd and 4th houses I smiled and shut off the nozzle and handed it to him and watched as he took off at full speed into the alley.

At the 50' mark the line became taut and he looked like a cartoon character as his feet came out from under him and he landed on his back with a big splash in a mud puddle. He jumped up and was screaming about not warning him he only had one section of hose and when he saw my big grin, he tucked his tail and disappeared behind the houses. I never had any problems with him again.

During the 1980's the City's Human Resources Department was pushing hard for more female firemen. I remember several hundred females taking the test with me at John Marshall High School back in 1981 but none were hired.

The older guys and the officers all said that they were unable to pass the physical agility test. The agility test simulated real life tasks that we performed at fires. In fact twice during the 80's I was selected with many others to go to drill school and go thru a simulated test as Chief Lewis and H.R. were locked in a battle of wills. H.R. wanted to lower the standards so as to hire more females. I remember thinking that no one should suffer discrimination, but that the test was fair and job related, and if you couldn't pass, too bad, as you would only be a danger to yourself and others on the street. We only had one female fireman and she had transferred in from the police Dept. and was well respected. In 1986 the city gave an entrance exam for the Fire Bureau. Once again no female scored high enough on the list to have any hope of being hired. One of the females on that list had a mother who worked for the school system and was attuned to H.R.'s desire for a weaker test. So she did what any good liberal would do, she sued the City. This lawsuit resulted in that test being thrown out and a different physical agility test given. The results have weakened the Fire Bureau and more importantly, put the lives of citizens at greater risk ever since.

January 2nd was our first workday in 1987. Lt. Val Jean was off on vacation and it was my turn to be in charge of the company as acting lieutenant. So many working fires and other weird stuff happened when I was acting that the guys labeled me a "black cloud." On New Years Day there had been a fire in a hotel in Puerto Rico that killed many people, so I had a hard time believing my ears when a well dressed black gentleman walked through our open garage doors at 0900 hours to calmly report that the Eggleston Hotel at 2nd & Leigh Streets was on fire. I walked out into Leigh Street and when I looked East, I couldn't see the intersection of 2nd & Leigh for the thick black smoke. The first thing I thought of was that the Eggleston was an 80 year-old three story wood frame hotel that had a restaurant on the first floor. The restaurant had a deep fat fryer that over the years had coated the entire interior of the building with cooking grease. In fact we all knew that although the building was a monument to the black history of Richmond, it was a deadly firetrap, a time bomb waiting to explode.

I sounded the house gong which summons all members to the apparatus floor and reported the fire on the direct phone line from the watch desk to the Fire alarm office. The operator picked up the phone and started to say "Happy New..." when I cut him off and told him to cut the bullshit and send a full first alarm assignment to 2nd & Leigh for a fire at the Eggleston Hotel.

Boarding the apparatus I figured there would be some sleeping people still inside the hotel and was struck by a sudden need to pray to God we would do our best to save as many as we could.

When we arrived at 2nd and Leigh we laid a hose line from the hydrant on the corner and went to the alley at the rear of the building as that's where the smoke was coming from. To my relief I could see that it wasn't the hotel burning but the two story house next door with the entire first floor going. Bryan had the nozzle and we met a severely burned man lying in the alley, his clothes burned completely off him. He told us another man was trapped upstairs. I radioed to the dispatcher to send us two ambulances. I looked inside the door and could see the stairs to the right engulfed in flames. I told Bryan as soon as he got water to knock the fire on the steps down and make a hard push into the first floor. I didn't have to explain to him I was going upstairs to look for the trapped man, as he already knew it. He also knew I was counting on him to put this fire out because I was going where even very few firemen dare to go, directly above and out of control fire. This was the "moment of truth firefighting" as life or death hung in the balance, including my own life. Donny Viar from 1 truck appeared in the alley with us just as Bryan's line was charged. As soon as Bryan knocked the flames down on the stairs I made a dash up the steps. Near the top I felt a hot wind descend on me like I had never felt before and had to jump up the last few steps. When I came down I landed in a pile of firewood that was stacked against the wall and me and the firewood went sprawling across the floor. Donny was below me and kept yelling up the stairs that they were relighting. I didn't care at that point because I had gone through the worst of the heat and conditions could only get better inside the bedrooms. Getting my feet under me again took a second because of all the firewood logs now spread out across the floor. I could feel my ears burning and my head got really hot as I went into the first room I found and closed the door behind me. Closing the door is an old trick to buy more time and to keep any more heat from entering the room you are searching. I made rapid sweeping motions as I crawled across the floor like a swimmer in a pool. A swimmer's pool is nice and cool while I was swimming in a room full of hot, black, carbon laden smoke that was a couple of hundred degrees.

After completing that room I went back out into the hallway and entered the next room. Using the same technique I quickly located the toilet by putting my gloved hand into it. During searches like this there is no visibility, everything is by feel, the smoke is that thick. The Third and final room was a little cooler as Bryan was making a positive impact on the fire downstairs. I searched everywhere but couldn't find the trapped man. The truck guys outside were breaking windows and the smoke was starting to lift just a little, Mike had joined me and I told him I couldn't find the guy. We both went back to the first room I had searched and with the improved visibility Mike located him between the wall and the mattress. We both grabbed him and hauled him down the stairs. When we got him outside we could see he wasn't breathing, but luckily the ambulance crew was waiting in the alley and took over for us. I went back upstairs and with visibility much improved I could see where his body was outlined by the soot.

During the search I had run my gloved hand across the top of the mattress and I figured I must have missed him by inches. It was a good lesson and it was something I always did, after the fire is out, go back and retrace you steps.

Later that night we found out that the guy I had spoken with in the alley had passed away from his burns, but the guy Mikey found upstairs was still alive and probably would survive. This job was so ironic sometimes, you mean the guy I talked to in the alley is dead, and the dead guy we grabbed from the second floor is alive! I said another prayer for the dead guy with a thank you to God for the fire not being in the Eggleston. My ears blistered up and I was starting to wonder if my skin would ever have a chance to heal all the way. I treated myself with the large jar of burn cream I had bummed from one of the EMT's from Commonwealth Ambulance Service.

The next workday Grayson was in charge and Gene-O was driving the wagon. At 0930 hours the alarm sounded for another first due fire at 1713 Roane Street with 5-14-46--51 for a building fire with a blind man trapped. I was on the backstep and we could see the loom-up of smoke from Leigh St. Gene-O was driving like his normal bat-out of hell self when we almost hit a car that pulled out in front of us.

The weather had turned much colder and snow was on the ground. When we turned into the block we could see a wood frame duplex with the entire first floor engulfed in flames, in fact there was so much fire the snow had melted at the front window and the grass was on fire. This was two days in a row with morning fires in a two story building with fire on the first floor with someone trapped on the second floor. Once again this job was so ironic. I had the nozzle and we pushed hard into the first floor. I could feel my ears burning again from the steam but that wasn't unusual. What bothered me was I was being burned on my shoulders where the air pack straps were. This was a super hot fire and I was probably pushing in too fast but the blind guy was somewhere in here. When we reached the rear of the building we went up the steps and a guy named Pat from 14 Engine and I searched for the blind guy.

Luckily for him he wasn't home because this fire was much hotter than the one on Second St. Once outside I went behind the hose wagon to take my coat off and check my burns. I was already blistered up pretty good on both shoulders, I had just put my fire coat back on when Chief Beaveman comes around the wagon and sees my burned ears. He summons an ambulance for me and I protest. "Chief these burns are from 2nd St." But he tells me I am going to the hospital. Now I am really pissed off because I have been burned many times in the past four and a half years. I had been inside collapsing buildings, I had fallen off loading docks, I had hurt myself everyway you can imagine, but I was proud of the fact I had never been taken to a hospital. To me it was a macho deal. I am a fireman; firemen get burns just like carpenters get splinters. If every time a carpenter gets a splinter he had to go to a hospital, houses would be unaffordable. But the chief insists. So I get into the ambulance for a ride to Stuart Circle Hospital.

Once at the hospital a team of doctors and nurses are checking my head and ears. They ask me if I am hurt anywhere else and I don't feel a need to disclose my shoulder burns. One of the blisters on my shoulder had already popped and the pus must have oozed thru my shirt because the next thing I know I was lying naked on the table. The doc nodded or gave the nurses some kind of signal and one nurse undid my belt and grabbed my trousers and underwear and in a nano-second I was fully exposed. I was quickly given an I.V. in each arm, all kinds of wires and oxygen and some other tests. Now I became concerned, was there something wrong with me? They were acting very concerned. These were only second degree burns on my ears and shoulders partial thickness burns that heal themselves in a couple of weeks, right? All I had to do was keep them clean right? Maybe I am really hurt? I kept asking questions but was told to stay quiet and keep still. The doctor took a pair of scissors and began to cut the blisters and loose skin off of my burns, it hurt like hell but I figured he was doing what was best for my injuries. When he finished I was bandaged up and looked like a damn mummy from the chest up. I asked the doctor if I could go back to work and he looked at me like I was crazy. "You're being admitted mister, and I am tempted to send you to M.C.V.'s burn unit." "Your right shoulder might need skin grafts and both of your ears may need reconstructive surgery." It seems that the burn crème I had on my ears from the 2nd St. fire had gotten so hot it actually burned me severely. He went ballistic when I told him I wasn't staying in the hospital. We had quite an argument and he called in the Fire bureau Safety officer and a Battalion Chief.

Both of them tried to keep me in the hospital but by that time my mind was made up I checked myself out of Stuart Circle against medical advice. The final battle was when they wanted me to come back the next day for dressing changes and further care. I informed them I wanted them to send my information to M.C.V.'s burn unit and I would be going to the experts on burns from here on out. I had never seen or heard of cutting burnt skin off with scissors, and I wasn't too happy about the lack of information I received, in fact they scared the hell out of me for about 30 minutes.

When I got home that afternoon I was sitting in my living room playing with my son and daughter. Jimmy was almost three years old and he was full of energy. Katie, my daughter, was five months old and loved to be held as she looked around and watched her brother bounce off the furniture. I had the fire radio scanner on for background noise and I turned my attention to it when I heard "5 engine on the scene 1202 North 1st St. we have fire out two windows on the top floor of a 10 story building, give me a second alarm to start off with."

"Damn," I thought, "I'm going to miss another good fire...but at least Grayson sounded nice and calm." "Yeah, we'll start of with a second," leaving the dispatcher to wonder if there will be more to come. That was Finner, always cool, I could see him getting out of the front seat of the wagon with that leather World War 1 bomber cap he always wore during cold weather and a big grin on his face as he leads the troops into the building.

I spent two weeks at home getting in Julie's way and feeling like a third wheel. But looking back it was a good two weeks as I got to spend real quality time with my family. I, like many young men trying to keep the wolf away from the door, spent a lot of time working. A 24-hour shift at the firehouse and going directly from the firehouse to building houses with Gary, or if it was raining, I would go to Eastern Motor Transport for a 10 to 12 hour day of staring at the back of a bulldog as I wheeled the Big Mack trucks full of gasoline through the streets of Virginia. My body healed up and I was pushing the doctor at M.C.V. to put me back to work. He finally relented and it was back to Leigh Street and my brother firemen.

It was a good thing I hit the books as hard as I did because two things happened over the next year and a half which changed the entire dynamic of the "B" shift. Mikey announced he was quitting Richmond and going to work for the Henrico Division of Fire located just North of Richmond. At the time the city lagged behind the counties as far as pay went, and Mike's father was an official of some sort in Henrico. We were really sorry to see him go as he had become one hell of a fireman.

In just three short years he went from being a guy who knew nothing about the fire service to someone who could be counted on no matter how bad conditions got.

The other was Grayson decided to transfer to 11 Engine on Church Hill. He had some friends at 11 and although they didn't run as much as we did at 5 Engine, They were a solid performer when it came to first due building fires. The glory days at Engine Company Number 5 were about to end.

- Chapter 19 -

People Trapped........Not

During my career as a fireman I noticed several truths. It seemed that when we were told that people were trapped inside a burning building, many times it wasn't true. There were dozens of fires that we went to that we pushed ourselves to super human efforts to find the trapped people. The burns and other injuries suffered by myself and my brothers were incredible. Of course some of the reports were true and resulted in a rescue from a horrible death.

The other thing was that it was common to arrive at a building fire (with no one saying anything) and during the advancement of the hoseline, to find someone lying on the floor unconscious. It was explained to me by the older guys that most people will see one serious building fire in their lifetime, and that event is so stressful they feel a need to be of help, they somehow convince themselves that an old lady, or a small child is trapped on the top floor. The other thing we encountered was the pet owner who escaped the fire and would grab you upon arrival to tell you "my baby is upstairs!" I quickly learned to ask "how old is your baby? Is it a boy or girl?" Many times the answer was "He's a four year old German Shepherd." Sorry lady, I love dogs and cats as much as the next guy, but I'm not going to take extraordinary chances for a pet, my wife and children need me too much for that.

The fire came in for the house with a 5 year old child trapped late one night. It was reported to be at 811 West Catherine Street. 5-10-41-52 was the companies on the get go. We had Henry Pollard from A-shift working with us that night. We traveled the four blocks quickly and arrived to find fire shooting 15' out the three front windows on the second floor of a two story wood frame house. Henry had the pipe and I was pulling hose for him. The front entry foyer on the ground level was also fully involved. We started to knock fire down from the front doorway when the electrical power line coming into the house burned in half and dropped onto the porch roof directly above us. The arc and electrical flashes looked like someone was welding directly above us. The sound and percussion wave in the air was also impressive. I entered the house in front of Henry as I have always had great respect for electricity. I was in the entry foyer when I looked up the stairs, I could see most of the fire on the second floor was towards the front of the house. If Henry knocked down a little more fire on the steps I thought I could make it to the second floor and search the rear. Henry was an aggressive fireman and he started up the steps with the line, I was right behind him when the steps collapsed and we ended up standing in a hole, chest deep, face to face being forced together by the debris. Henry tried to tell me something, but I didn't understand him when he grabbed me by the coat and picked me up and set me down on the portion of the steps still standing. I asked Henry if he was ok and he told me "Now I am." It seems I had pinched his low pressure hose for his air-pack and he couldn't get any air, hence the lift up and out of the hole. I climbed up the steps and started to search for the 5 year old boy. After covering two rooms I met up with Buzzy from 3 truck who had come in over a ladder. "Did you find anybody yet?" He asked me. "No Buzzy I haven't." I replied and we teamed up and worked together and searched the rest of the house. It was another case of reports of people trapped who aren't.

The lieutenant's test was given during this time and I ended up in the top 10 candidates for promotion. My first promotional interview was on September 18, 1987, my 25th Birthday. The City of Richmond had a "five plus the number of vacancies rule." If they were promoting two people, the top seven from the list got an interview. This gave the Chief a chance to pick the best candidates for the job. I had a good interview with Chief Lewis but a week later when I received my letter he advised me that I hadn't been promoted. What bothered me was being skipped for two men who I knew, and didn't think they were better than me. One of them also happened to be black.

In 1983 The City of Richmond implemented an affirmative action policy that caused race to be considered in all hiring and promotional decisions. I had heard the older guys complain since I'd been hired about the affirmative action policy but until now believed they might have racist tendencies. Here I was a decorated veteran in the busiest Engine Company in the city, being passed over for a black guy who was known to be a coward. It didn't sit well with me but I decided to let it pass without complaint. Now the unfairness of how my city promoted was in my main focus. I watched closely the guys who were awarded their promotions thru affirmative action.

I was on a fire one night with 14 Engine, we were picking up hose when a newly promoted black lieutenant insisted we pack a preconnect the way he wanted it packed. I looked at Pat from 14 and he nodded to me that he knew it was wrong too. We packed it the way this affirmative action boss wanted it. The next work day I asked Pat if he fixed the A.A. boss's mistake? He told me once they got back to the firehouse, they waited till the boss had gone to bed and when the pulled the 200' preconnect hoseline, they found 150' packed in a circle……..connected to itself, and the remaining 50' from the pump discharge to a nozzle.

Another time we were sent to city hall for an electrical fire and we arrived in the sub-basement parking deck to find an electrical transformer the size of large dumpster smoking; the oil inside the thing was cooking and the paint was melting and running down the side of the metal cabinet, and it was making thumping sounds from the boiling oil inside. This transformer was surrounded with a cinder block wall and another newly minted black leader had us standing in the enclosure near this ticking time bomb. I informed him I didn't think it was too smart to be that close to a huge electrical transformer that was melting before our eyes. The lt. took offense and told me he was in charge and that everything was o.k. I walked away as did Finner and several others and we put some distance between us and the impending disaster. Grayson and I discussed whether we should even try to assist our A.A. boss when the metal cabinet ruptured and started spewing hot oil on the clueless leader.

A few minutes later a power company lineman came up to the enclosure and took one look at the transformer and spun on one heel and quickly ran out of the building. A.A. Boss sees this and finally figures out it might be dangerous and comes over to where Finner and I are standing but isn't man enough to acknowledge his mistake. I was trying not to become bitter and I had worked for several great black officers but even Ray Charles could see that the push to promote as many blacks as possible was quickly becoming unsafe for both firemen and the citizens we served. I remember one officer rationalizing the problem by claiming that the poorly prepared men will grow into the job.

Sometime these guys were downright comical and embarrassing at the same time. At another job we found a late model Mercedes Benz with fire in the engine compartment, the main wiring harness had short circuited and was burning. A man in a very expensive business suit stood-by watching silently while we extinguished his automobile. After the fire was out our "distinguished" leader called the man in the suit over to the car and explained that, "Your car been burned-up cause you gots a shortage here." The man looked at this goof ball officer and said, "Lieutenant, there is plenty of wire there, I think you mean a short circuit caused the fire." To which our boss replied, "Yeah, that's what I told ya!" The man looked at us and just shook his head in disbelief and Finner in his normal break-the-ice humor tells the business man, "Don't look at us, we gotta work for this guy."

It was very difficult to take these officers seriously, and some of them I had worked with at fires over the years, and knew they weren't very bright. For instance one A.A. officer as a fireman had helped cut a ventilation hole in the roof of an apartment building fire one night and then proceeded to fall into his own hole. Fireman Skylight as he was then nicknamed, was working overtime with us at 5 Engine one afternoon as we pulled up to a room and contents fire on Charity Street. "Skylight" had laid-out and instead of getting back on the wagon he stood by the hydrant while I positioned the pumper to hook up the soft suction hose on the passenger side at First and Charity.

Using the soft suction hose on the passenger's side was a little trickier than the driver's side but luckily we did it enough that you developed a technique. Mine was to get the distance from the hydrant just right and then watch thru the jump seat window until the hydrant disappeared from view, this usually provided for a perfect kink free hookup. I put the engine in pump gear and grabbed the two and a half inch supply line and connected it to the discharge on the driver's side pump panel. As soon as I had connected the supply line the wagon driver called to "Let the water go!" our term for "fill the supply line with water." I did so using booster tank water. As I was adjusting the throttle to regulate the pressure, "Skylight" comes running around the big Mack pumper to report that "The hose wouldn't fit the hydrant." I quickly ran around to the hydrant because I only had five minutes of water in the booster tank. It is always dangerous to run out of water at a fire because the guys inside are deep in the building, and depending on that water. I was also worried that I had made a mistake and positioned the pumper too far from the hydrant. I could fix that problem with two fifty foot sections of two and a half but would incur the wrath of Captain Emerson afterwards because the only correct hook up for the pumper of a two piece engine company is the soft suction hose. I could already hear him ranting that this is the only method that ensures maximum water from the fire hydrant. I grabbed the five inch diameter suction hose with the big brass coupling with large lugs to provide a means to tighten the connection. Leaning into the hose and using my body weight I began to carefully start the threads to make the connection. Every time I thought I had the threads started "Skylight" would reach around the hydrant and try to help, each time it would cause a cross thread. I had already lost patience with Skylight and cursed at him and told him to keep his hands to himself. I finally got the connection made and angrily asked him to turn the water on. I returned to the pump panel on the driver's side and could see the intake pressure gauge jump-up as he turned the hydrant on.

He then came around and asked me what he should do. I went berserk, was he stupid or lazy? I strongly suggested he get up the street and help the real men doing the work. After the fire this same guy is standing around talking while the rest of us were draining and packing hose. I was getting angry again when he informs me that I hurt his feeling by cursing him at the hydrant. I exploded and went after him, I was going to kick is ass right in the middle of Charity Street, but several of the guys grabbed me and wouldn't let me go until I calmed down. I called him everything but a man and informed him that when he worked with us, he was going to pull his fair share of the weight. He was promoted to Lt. a couple of years later and was what the Army called "danger close." Skylight was the type of guy to watch out for, as he was a train wreck about to happen any second.

I have been accused many times of being insensitive to the feelings of black people, usually by someone with nefarious intentions, I never took those accusations too seriously as I rescued 23 people from burning buildings in my career, and 22 of those people were black. I assisted too many young black men who had been shot and everything from car wrecks to patching up the loser of a fight. You name it the local firemen are usually the glue that holds poor neighborhoods together. The real irony of the community organizers who were hell-bent to secure jobs for people who were unqualified for them was that they were the same demographic most likely to suffer a serious fire. Let's face it; most fires don't happen in rich white neighborhoods. The fact of the matter is the busiest fire companies in the United States are all located in areas that have four things in common.

1) High population density; men, women, & children cause most fires.

2) Older buildings; old structures require more upkeep. That costs money.

3) Poor people; people living in poverty can't afford to properly maintain heating systems, and other important building features like smoke alarms.

4) Black people; this is one of the biggest tragedies in the U.S. today.

I have studied these cultural phenomena, and don't fully understand them.

My friend Pete Lund, who worked in New York City, told me of an area in Brooklyn that had been burning for years. Serious fires occurred on a regular basis. Then in a short period of time Hispanic people moved in and the blacks moved out. Pete explained that it was like a faucet had been turned off. The serious building fires stopped! The first three items remained. The area was still highly populated, the old buildings still stood, (what was left of them), and the residents were still poor. So why is it, that neighborhoods with lots of black folks burn so frequently? After many years of study I have a few theories, but that's a topic that would fill another book.

So there you have it, you can't always believe everything people tell you, just because they tell someone's trapped in a burning building. It may be true or it may be a physiological need, to feel like they can help out. If someone calls you a racist or some other bad thing, they may be motivated by some other reason. Who knows what it is, I never spent much time worrying about it. I learned from my grandmother at a very young age that, "sticks and stones will break bones but words will never harm you". But you still must get in there and make your search and do your best even for those who hate your guts and don't understand why.

Sounds like something I remember hearing in church; whatsoever you do to the least of my brothers that you do unto me.

- Chapter 20 -

Lieutenants Bars, Are They Worth It?

My first taste of Affirmative Action left a bad taste in my mouth as I suddenly realized that what the older guys had been telling me for years about the promotion system being unfair just might be true. As I said before I also started taking notice of the other young black officers. I noticed some were excellent leaders and I also noticed some white officers who were also clueless idiots. Many of the less than stellar white officers had been promoted in 1970 when the city annexed 23 sq. miles of Chesterfield County.

In 1970 The Fire Bureau added four Engine Companies and two Hook & Ladder Companies to protect the "annexed area" as it was referred to. This created six new captains positions and twelve Lieutenants jobs. In the firehouse kitchen when the subject of A.A. came up, many black firemen would point out look at Lieutenant So and So, or captain What's-his-face as examples that the best man didn't always get the job. And I could see that some of these guys they pointed out were not very good leaders. But I figured it was more of a function of having 18 new promotions all at once, plus the normal retirements and other vacancy's that caused a few bad leaders to slip through the system.

In September of 1988 I sat for my second interview with Chief Lewis. A full year had passed and only five lieutenant's positions had opened up: ten eager firemen going out for only five spots. Promotional opportunities were that few and far between since age 60 was the normal retirement age and you literally had to wait for someone to retire, or die, no one ever seemed to quit.

Chief Lewis was an elegant speaker and a pleasure to interview with, as his questions were probing but they way he asked them made me feel at ease.

The Richmond Fire Bureau was a small community and almost everyone knew everyone else. Secrets were hard to keep and he had heard of my disappointment in his last round of promotions and he asked me about it. We talked at length about Affirmative Action and he had a different perspective than I did. I remember trying to make my point that if discrimination was wrong in the 1950's, it is still wrong today. It was an honest, mature and civil discussion but he let me know that he was the boss and would promote as he saw fit. He also ended by suggesting that if I was serious about being promoted, that I should look into attending a local community college to attend Fire Science classes.

A couple of days later the promotion list was released, once again 3 of the 5 new lieutenants had scored lower than me on the test were now getting promoted.

I thought long and hard about what Chief Lewis had suggested. It was a hard decision as I had always hated school, but I was also a little older and wiser now and soon enrolled at J. Sgt. Reynolds Community College. I dove into classes there with the same intensity I had at a room and contents fire. It meant even less time at home with Julie, and my children, Jimmy, & Katie, but the gauntlet had been laid down and I was never one to back away from a challenge.

Over the next couple of years I was taking 2 classes per semester while working at the firehouse, and working a part time job as much as I could. The promotion list expired after one year but had a clause that the city could extend it another twelve months, which they always did.

The list I was on was two years old and it died while I was in the number one spot having been skipped six times. That fact angered me but in spite of this, soon I was on the deans list at Reynolds Comm. College. I found learning to be great fun and while the term papers were labor intensive, they also taught me the lesson that the harder you work at something, the more you learned. I did term papers on the subject of high rise building fires, residential fatality fires, etc. My life long passion of being a fireman was really coming into focus as the results of my research usually confirmed what I had seen on the streets of Prince George's County, Maryland and in Richmond. It also taught me the value of critical analysis. Critical analysis and the ability to think for yourself and not just follow the crowd, it is an essential skill for a real leader.

By mid 1988 both Grayson, and Mikey had left us for greener pastures. Our two new rookies assigned to 5 Engine had never been in the fire service before, but both were politically connected. Trina was hired after she failed the entrance exam and her mother sued the city to have the test standards lowered.

It was apparent from early on that this woman was a one person wrecking crew. She was the 2nd female hired in the Fire Bureau and was always seeking to be the center of attention. Isaac was the nephew of a Battalion Chief and was quick to remind anyone that "My Uncle be the Chief". It wasn't long until he was nicknamed Munclebe, pronounced "Mun-cul-be". Both were very slow at learning their job responsibilities and it made firehouse life very interesting.

Trina's first fire came in about 0900 hours on her second day on duty. It was reported as an apartment fire in the 800 block of Bowe Street with children trapped. Gene-O was in charge and I was riding in the jump seat area with Trina and we could see heavy smoke from Leigh St. I told Trina to make sure she had all her gear on and to follow the hoseline when she got ready. We arrived first in on side one with a two story type 3 (Brick & joist) quadraplex with two rooms on fire in the bottom right unit. I pulled the inch and a half cross lay to the front door and entered. The floor plan consisted of a living room, bedroom, and a kitchen with a small bath in the corner.

The apartment was known as a railroad flat as you traveled from room to room because it had no hallway. I crawled to the bedroom doorway and encountered heavy smoke conditions and two fully involved rooms in front of me.

I opened the pipe and deflected the water off the ceiling with a straight stream pattern and the first room darkened down. I crawled about four feet into the room and ran into a queen sized bed. I was sweeping my left hand under the bed searching for anyone who might be in this apartment when the room reignited. A blast of heat hit me and I was surrounded by orange flame. In less than a second I knew I was in serious trouble. I raised the pipe up to cool off the environment when the sudden urge hit me to get up and run, I was being burned and burned badly and I could smell my own skin burning even with the face piece of the air-pack in place. I gritted my teeth and quickly shuffled backwards on my knees and luckily went right through the doorway to the living room and stayed at that safe harbor for the next five minutes flowing water into the now darkened room. I was always taught not to flow water into smoke, but I was that scared and I knew I was badly burned so that it took five minutes before I could summon enough courage to crawl forward again.

Trina arrived behind me and kept hitting me on the shoulders to let me know she was there, the problem was every time she touched me I was getting burned, my gear was that damned hot. Slowly we advanced into the bedroom and with water flowing continuously we arrived at the doorway to the kitchen and encountered the crew from Number 6 Engine Company. They had arrived after I entered the front door and stretched a two and a half inch line to the rear of the building where the rear wooden porches were burning. After extinguishing the porches, they entered the back door. The more powerful 2 ½" line had created enough force to push the fire right overtop on me. I knew it wasn't done on purpose and is one of those friendly fire type things that sometimes happen in the heat of the moment.

I ducked out the backdoor to check my burns and ran into Radar from 6; he stopped and gave me a look like I was an alien creature. Once outside I could see something hanging directly in front of my face about ten inches in front of my mask. I doffed the equipment and what I saw caused my great concern. The hard plastic "Philadelphia" style helmet was badly deformed and the clear eye shield had shriveled up and was now just a blob of melted goo hanging in front. My S.C.B.A. face piece had also began to melt and I quickly realized that had that piece of equipment failed, one breath of superheated air would have been my last. I had some pretty serious burns to my head, face, neck and shoulders that resulted in another trip to the burn center.

It was a couple of weeks later when I returned to work to find out that the Bowe street fire had taken on a life of its own. The fire chief's office had inspected my gear and determined that my helmet had been exposed to temperatures of up to 1500 degrees They also discovered that my Nomex hood was ruined and the vapor barrier in my turnout coat was destroyed in the back and shoulder areas.

Chief Mosley requested I write a statement of what happened. I was happy to comply until I discovered he was going to use the statement to charge Captain Emerson with some sort of trumped up discipline. I knew in my heart that Captain Emerson was a good man and it was an honest mistake. I also made up my mind not to be a part of it. I wrote a letter explaining what I saw and did with the exception of meeting Number 6 in the kitchen, I also used it as a tool to praise Chief Lewis for buying us the best equipment on the market and how it saved my life. The letter angered Chief Mosley but there wasn't much he could do about it as I was the only one on 5's hose line.

Gene-O was laying at the front door with-out an airpack and Trina being brand new was understandably slow in getting into the building. The Division Chief came to 5 Engine a couple of days later and he wanted a letter addressed to him about what happened on Bowe St. Again I didn't give them anything that could harm a damn good man, and an excellent fire ground leader. I couldn't believe it; they were trying to get me to say something bad about a man who was like a second father to me. I remember looking at this Division Chief and thinking If Captain Emerson asked me to snap your neck, I'd do it without a second thought. This Chief was a known political hack and they were trying to punish a real leader of men. Finally Chief Lewis himself came to 5 Engine and wanted to talk to me, not a formal talk but he actually Told Lt. Bridges he wanted to talk to me alone at the kitchen table. He started off asking about Bowe St and how concerned for my safety he was, he told me about a guy he worked with in the Philadelphia Fire Dept. and how I reminded him of his old friend. He told me that he considered his old friend one of the best firemen he'd ever worked with and how he was killed in the line of duty. I was both humbled and puzzled; did he really think I was trying to kill myself? Was this a way to soften me up to spill the beans on the good Captain? Or was he really concerned for my wellbeing? I knew that I believed deeply in what I was doing, I believed I was doing the work I was put on Earth for. And I knew that if someone was trapped in a burning building, we were their last hope. I thanked him for taking the time to speak to me and promised to be careful, but, I believed in God and I knew he had a plan for each of us. The Fire Chief then examined my still raw neck and admonished me for returning to work so soon, but I could sense a deep admiration for a job well done.

The story of Bowe St. was told over and over again at events like the annual Credit Union Dinner and other places where firemen congregate. Captain Emerson was never charged with any infraction but it was another lesson of how the politically ambitious would try to step on the neck of anyone viewed as a rival for things like promotions. It was also the start of a deep dislike for anyone who took the oath to protect life and property and wore the same uniform as I did, but was nothing more than a coward in the face of danger. And the older I got and the more I looked around, I saw that there were plenty of these types of characters.

The lieutenant's list expired and it was time to study for the next one. Every moment of every day that I wasn't busy, my nose was buried in a book. Julie's family was from New York City and we went to a family reunion in the Bronx. Her uncle didn't understand why whenever nothing was going on I was in my study material and he gave me hell about it all weekend. It must have worked as I finished third on the lieutenant's test.

Captain Crick was now the boss at 5 Engine and he wrote up a nice letter of recommendation for me. He also wanted me to relinquish the nozzle to Trina and Munclebe as he was concerned that after I was promoted he would only have one experienced man, Bryan, on that shift. I was all for helping to teach the rookies, but we soon found some interesting attitudes that would make the next year very interesting. Every new fireman in Richmond is tested at 6 months and 12 months on the job. These tests confirm that you have learned all the street and hydrant locations in your first alarm area. Every time we tried to quiz Trina on streets she had some lame excuse for not knowing the information. Lt. Val Jean even got involved but she was a hopeless case. Whenever things got uncomfortable for her she would accuse the white guys of being racist and the black guys of being "Uncle Toms." It got to the point where many guys just steered clear of her. I asked her once about how she planned on passing her six month test, she looked at me and said, "They wouldn't fire me, I'm a black female." She quickly set the stage where even most of the officers were afraid of her. I heard time after time that I too should steer clear of her because she was a one person train wreck, but I thought that talk was nonsense. We had to work with her and I figured we could make her conform to the will of the group. I felt that letting her get away with this bad attitude was bigotry in itself. She wanted equal opportunity; I was willing to give it to her.

Sunday April 2nd 1989 was a warm spring day at 5 Engine. Sundays were usually quiet but fires come in at anytime day or night. A little after 5 o'clock in the afternoon, a full first alarm assignment was sent to 500 South 14th Street for fire in a warehouse. 13 Engine went on scene with fire in the building, and a group of us congregated around the watch desk to listen to the fire. A large computer print-out book at the desk indicated that we were 2nd due on a second alarm and this sounded like a big fire. You could hear the excitement in the voices barking out orders on the radio and soon enough Battalion Chief Griggs requested the 2nd alarm. Traffic was light and we made good time enroute to the fire we would spend the rest of the 24 hour shift at. We took a hydrant at 14th and Dock Sts. and laid all of our hose just to reach the front of the building. 13's wagon was in front of the fire and we borrowed 400' of hose from them to supply our stang nozzle. Captain Crick wanted to set up the stang in front of a loading dock to shoot water into the large overhead doors.

Two 2 ½" hose lines disappeared into the smoke and I convinced him to take the stang nozzle into the building to support the brothers manning those two big hose lines. Both Engines 6 and 33 were inside and had taken a position between wings of the building and were struggling to keep the fire from spreading into the unburned wing. We set up our portable water cannon capable of flowing 1000 gallons per minute and soon were beating the large stacks of burning paper.

The building was approximately 1500'x350' and was full of recycled paper bailed and stacked 15' high. The building itself was type 2 or, non-combustible steel. The problem here was the fire was so hot and the lightweight steel was unprotected, and the building was collapsing above the fire. The fire science classes I was taking gave me an understanding of how this type structure would behave under these conditions and we kept the powerful stream moving across the ceiling to keep the steel cool and it acted as a deflector to bank the water onto the fire. Four air cylinders, and almost two hours later, the fire was still burning under the collapsed roof but the area we could hit with our stream continued to stand.

The Chief ordered all companies to evacuate the building, but we didn't want to leave, up to this point we had kept the fire out of that wing and fought too hard to just give up. But the Chief prevailed and we were soon disassembling our stang gun. The area in front of the loading dock where we pulled our hose over was now a 3 foot deep lake and I told Bryan to wait as I got down into "Lake Griggs" as we called it and had him hand me the parts of the stang. I took them about 50' from the building and set the down on dry land. As we were doing this that political minded Division Chief shows up and is barking out orders and asking questions. I am so tired I have to work at standing up when this deep booming voice I know to be Bryan's announces from behind the wall of smoke, "Hey Chief, FUCK YOU!" The Chief looks at me and asks, "Who said that?" When Bryan again says "I SAID FUCK YOU!" The Chief gets a quizzical look on his face and goes on to the next group of fireman to assign some inane task. The rest of us look at each other and bust out laughing.

After we got everything out of the building and set up in the parking lot, we threw water at this lost cause for the remainder of the night. The next morning we were relieved on scene and it would be 3 days before the last fire company left the scene.

Throughout the summer we took in plenty of working fires but it seemed as though I would end up with the nozzle even though I was trying to give it to the rookies. Captain Crick was getting impatient with me, but it seemed that I was always winding up on the pipe.

Jake Rixner

Preston Street was a classic example. The fire was in the 2nd floor rear bedroom in a Century old two story row frame. Four houses shared a common roof and we had fire in a middle unit. This means the fire once it spreads to the attic area, is going to burn the roof off of its three neighbors houses. Trina had the pipe and I was backing her up. Munclebe had recently started driving and was the wagon driver; he was having trouble understanding how to transition from the booster tank to a hydrant water supply, and had a bad habit of closing the booster tank valve before he opened his intake valve to obtain the water being pumped from the hydrant. This resulted in a loss of water for a few seconds and no matter how many times we went out and practiced, he just couldn't get it. I assisted Trina in advancing our line to the top of the stairs; the fire was in the room to our left and once we got water, she began to hit it from the stairs. I heard the tell-tale change in R.P.M.'s of the diesel engine indicating that we lost water when the hose went limp. Trina didn't shut the nozzle off and was talking to me when Munclebe opened his intake valve and water raced to the nozzle which Trina wasn't holding tightly and the brass pipe quickly struck her head about ten times before she pushed it away and dove down the steps past me. I crawled up the line and retrieved the out of control hoseline before it split open somebody's head like a melon, and quickly knocked down the fire.

Captain Crick heard that I had the pipe on Preston St. but not the rest of the story and was angry with me when he ordered me not to touch the nozzle, but to help the rookies. I couldn't help but smile at him which made him madder. I promised him that I would not touch the nozzle again. I knew things were about to get interesting.

The fire was reported at 34 West Jackson Street. The corner of Price and Jackson was one of seven corners in our entire first-due that didn't have a hydrant. Munclebe was driving and went the quickest way, which unfortunately left us with-out a water supply. (We were running as a triple combination pumper, i.e. one piece.) Flames were 5' out the window of the front bedroom in a 2 story wood framed duplex (two houses under one roof). I knew if I went upstairs with Gene-O and Trina, I would end up violating the Captain's orders so I stayed at the bottom of the stairs. I was feeding them hose when the line stopped moving forward. I could hear an argument between Trina and Gene-Owhen Lt. Val Gene ordered everyone out of the building. Returning to the street I expected to see fire through the roof, but the fire was still burning out the front window like it was when we got there.

When Battalion Chief 1 arrived he quickly sent everyone back into the building. Radar and Mike Hale from 6 Engine quickly extinguished the fire while Trina and Gene-O goofed around. I was mad enough to explode but was trying to remain calm. I asked our Lt. why he pulled us out of the building. He replied that he expected to see the fire knocked down in a short amount of time and when the fire continued to burn he figured something was wrong and gave the evacuation order. I asked Trina what happened upstairs and she told me it got real hot and she stopped moving forward, then she added that she didn't have a death wish like I did. That pissed me off. Death wish? She was goofing around and was probably less than 5 feet of seeing the fire and putting it out. The easiest building fire to extinguish is one that's venting out a window when you arrive. Most of the heat and dangerous gasses are being channeled away from the nozzle. No, this ignorant young lady had put her own life at risk by stopping.

I asked Gene-O what happened and he said Trina just stopped moving forward. I told him he should have tried to help encourage her or take the pipe and finish the job. So a simple room and contents fire was put out by Number 6 Engine Company, Thanks Captain.

I could fill several more chapters in this book of the idiotic behavior of Trina and the total destruction of a once sharp firefighting team but I don't like to feel sorry for myself for too long. Heck, I was about to be promoted. There was a small problem though. Trina went to Chief Lewis's office to report that, "all the white guys are picking on me." She wanted a transfer to a slower unit but Chief Lewis launched a full investigation, after all we can't have racist white guys making poor black women feel uncomfortable.

I got a phone call from a secretary at Chief Lewis's office telling me that Bryan and I were being transferred to slow Engine houses where the Chief knew we would be miserable. She also told me that Chief Lewis stated that I would never be promoted.

I went to see a lawyer who after I presented my case informed me that he knew Chief Lewis, and that Chief Lewis wouldn't do anything that stupid and that maybe it was me that had a problem. I left his office upset as he was the best lawyer in Richmond and he cut my knees out from under me.

Two days later a transfer list came out. All the white guys on our shift were transferred, Me to 19 Engine which averaged less than 1 run per day, and Bryan to 25 which barely broke 200 runs a year. All the officers were lifted too. I called this lawyer back and informed him what Chief Lewis had done, he told me to come back down to see him and bring $200.00 dollars with me as I was now a client.

At the lawyers office we went through all of the things that had happened over the past two years. I learned that I had to exhaust the grievance process before I could file a law suit and that black females enjoy a lot of protection under the law, but they were not protected from not knowing their job. In short if I could prove incompetence, I could win a law suit. Incompetence, hell I could write a book about this individual. I went to our log book and through fires and other calls I wrote a ten page list of mistakes and other events that indicated that this person had little respect for the duties she held as a fireman for the City of Richmond.

I filed my grievance against Chief Lewis and had a hearing with him. I was working at 19 Engine near the University of Richmond when Captain Bernard Emerson told me that Chief Taylor was coming to pick me up for my meeting with the Chief. He was the brother of my first captain and was once at 10 Engine, I could tell he didn't like the controversy. Chief Taylor arrived and tried to intimidate my by announcing that Chief Lewis wanted to talk to me. I was pissed off but smiled at this political hack and informed him that yeah, he was going to meet with me, but then again he didn't have a choice. Chief Taylor drove towards the Chief's office and was coaching me on how to act. Be respectful, remember the respect for the office of the fire chief, and so on. I listened to this as long as I could stand it.

We were stopped at the traffic signal at 8th & Main when I unloaded on this poser, "Hey Chief, thanks for your advice, but you gotta know that if I see you on the street off duty, one of us is going to the hospital because I am going to beat your ass." I meant every word of it. This was the same guy who tried to write up Richard Emerson and would step on his mother's neck to get promoted. There was no love here, when we got to the Chiefs office he quickly went his own way.

I sat in the entrance foyer for an hour and a half before Chief Lewis arrived and informed me that "he would see me now". The 90 minute wait had me hot under the collar. Once in his office he went over the entire case and informed me that he was the fire chief and I would work where he wanted me too. He denied ever making the statement that I would never be promoted and Bryan and I would work in slow houses where we would be miserable. After letting me know that he was the boss he put my grievance in front of me and wanted me to sign off on it. I couldn't believe it. Here we were, two men with different perspectives. Did this guy really think I was that dumb? I refused to sign off settling the grievance, the next step was the city managers office, his boss, and I looked forward to explain how he treated his employees.

Chief Lewis lost his composure and started talking street lingo asking me what I wanted. Not knowing if the office was tapped I grabbed my shirt collar. "So that's what this is about." Chief Lewis said. "No Chief not at all," I replied, "Just sit back and wait, Trina will prove my point in due time".

The summer of 1990 was a long one as 19 Engine Company was slow, so slow it took 3 hours to watch the television program 60 minutes. Captain Emerson was like a second father but I was depressed. For instance I drove an oil delivery truck for Woodfin oil in this neighborhood I knew the streets like the back of my hand. When a run did come in and I was driving, and me and the Captain would be in the front seats and look back to see the other two guys studying the map on the wall. More than once I asked the Captain to leave the clowns at the map. He never let me do it but I wanted to so badly. After nine months of torture I was summoned for a Lieutenant's interview in late September.

Chief Lewis was polite and respectful, but he knew my intention was to win my lawsuit. Our interview was cordial and he informed me that I was correct about Trina. She was facing charges for performing oral sex on Munclebe on the Engine. I just smiled and was glad he realized that Bryan and I weren't racists. He asked me where do you want to work lieutenant? I smiled and told him 3 TRUCK COMPANY. "3 Truck huh," and he rubbed his chin. The interview soon ended and I knew I was being promoted.

A couple of days later I received a letter from the Chief's office. It stated I was being promoted and sent to the number Two Truck Company in Church Hill. Damn, I was going into the war zone with some of the best firemen in the city. At the promotion ceremony I thanked him and told him I was dropping my lawsuit. I was back on top of the game and now I had a chance to make things better.

- Chapter 21 -

Promotion Day

October 5, 1990 was promotion day. Four other firemen and myself were about to become the leaders of men. It was a step I took very seriously. I had been a professional fireman for eight years and up until now I had basically only been responsible for myself. From this day forward I would have the lives and well being of other men in my charge. Some call this the burden of command. My mother, grandmother, wife and children all accompanied me to the Fire Chief's office to pin the gold badge on.

It was a bright and sunny day as we arrived at the Chief's office. As soon as I entered, I knew this wasn't going to be a good experience. The Fire Chief's office had a reputation of being like a shark tank. People that worked there and those that hung out there were the cut throat types. The kind of people who would throw somebody under a bus if they thought it would help their career.

The first person I ran into was Skylight Smith. He wasn't getting promoted but he was at the office brown-nosing. Now here was a guy they skipped me to promote, but he didn't know which end of the hose the water comes out of. Skylight rushes up to me to congratulate me on my promotion and immediately offends me. I look him straight in the eye and tell him to get the hell away from me.

Chief Lewis is behind me and takes offense at my statement. I just can't tolerate false emotions. If I don't like somebody, I can't pretend to like them in situations like this. I guess I wouldn't make a good diplomat. Chief Lewis invites me into his office and asks me what my beef is with Skylight? I explain that he isn't a fireman and I consider him to be stealing money from the city every time he takes a paycheck. Chief Lewis gives me a pep talk about understanding different types of people and asks me if I am going to be a problem lieutenant. I laugh and tell him he wouldn't have any trouble from me as long as those guys at 2 Truck are good firemen.

Chief Beaverman is there and he teases my children. He shakes my hand and congratulates me and instantly I am at ease. We take the oath of office and Julie comes up and takes off my silver Fireman's badge, and pins a bright shiny gold lieutenant's badge on me. Photos are taken and then we are dismissed.

On the way home my grandmother asks me to see my new firehouse. As we drive up Broad Street towards 1 Engine and 2 Truck's fire house She asks me if my new job is safer than my old one and without thinking I explain that doing my old job I had to drive the fire engine once every four days and would miss out on interior firefighting but that now as a lieutenant I would get to go into every burning building from now on. This disturbs her and I instantly regret telling her the truth. I should have shielded her from the realities of this job. I was never one to talk shop around my family; I wanted to shield them from the danger and worries. But around my buddies in the fire service, I was 100% into the job. My wife learned more by eavesdropping on my conversations with them and would later chastise me for not telling her about my work.

Once again the weekend ends too soon and soon Julie, the children and I are bidding farewell to our visitors. Later, as I lay in bed my mind wanders as I wait for sleep to overtake me. Like a slide show I can see Donny Viar dropping in front of me in the collapse on Marshall St. The guy who came out of the flames on Catherine St to report "Willie's in the back man". Mom Mary screaming "Hey Motherfucker" at 2nd & Leigh Sts. Every time we passed. Grayson saying "Yessim boss" to our supervisors. Bryan doing his best Monty python impressions. Mikey clucking like a chicken saying Paw Paw… …., Paw Paw! Gene-O hitting on some woman on the front ramp of the firehouse and spinning around and walking smack dab into the flagpole. Lt. Pulliam with his pants and underwear around his ankles at 10 A.M. at Hermitage & Leigh Streets yelling for me to "come here!"

The time we almost ran into 1's wagon at 14th & Franklin Streets, we had discovered a new way to beat everyone into the James Monroe building at 14th & Franklin. It involved getting onto Interstate 95 directly behind the firehouse. The exit ramp for Franklin St. dumps you right in front of the Building but is a short ramp. Mikey was driving and didn't know that and he locked his brakes up and slid through the red signal in front of #1. Thank God Engine 1's wagon driver saw us coming and stopped. Their lieutenant was a true southern gentleman and I had never heard him cuss before that day: the blank stare in Mrs. Brown's eyes as Captain Emerson and I carried her away from the flames: the good times with the men in the firehouse kitchen drinking coffee, playing dominoes, and giving each other a hard time.

As my mind wandered as I waited for sleep to come, I said a silent prayer of thanks. Thanks for letting me know what my purpose was on this Earth. And I prayed that others find their niche. Thanks for my health and for my family. Thanks for all God's blessings and one final request, Please help me to become the best fire officer I can be. And then I had to smile to myself thinking, here we go again.